T0215333

Critiques of Knowing

'Reading *Critiques of Knowing* is like looking into a kaleidoscope, where bright fragments are set in mirrors and new and beautiful patterns form. One of my special pleasures in reading her are the delicious asides; these are very light, very fast, Lynette Hunter makes some devastating theoretical reflections, and the text races down the page.'

Hilary Rose, author of *Love, Power and Knowledge*

'*Critiques of Knowing* is a book of extraordinary intellectual breadth and depth....I know of no other book which brings together this astonishing range of currents to weave together a critique of such a large part of our contemporary intellectual culture.'

Alison Adam, author of *Artificial Knowing*

'*Critiques of Knowing* is a highly original text....I know of no other book of feminist epistemology which spans such wide-ranging disciplines.'

Gill Kirkup, The Open University

'This is a book that is thoroughly engaged with on-going debates in a range of interconnected topics....It makes a distinctive contribution to current debates and will, no doubt, generate further debate.'

Sarah Hutton, President of the British Society for the History of Philosophy

'This erudite, elegantly written book engages provocatively with questions at the forefront of scholarship in the humanities at the end of the twentieth century.'

Lorraine Code, York University

Lynette Hunter is Professor of the History of Rhetoric at the University of Leeds. She worked for a number of years as a biochemist and laboratory technician, and has recently become widely known for her performance art lectures. Hunter has published in the fields of critical theory, feminism, literary criticism, rhetoric and the history and philosophy of science. Her books include *Rhetorical Space, Modern Allegory and Fantasy, Outsider Notes: Feminist Approaches to Ideology, Writers/Readers and Publishing*.

Critiques of Knowing

Situated textualities in science, computing and the arts

Lynette Hunter

London and New York

First published 1999
by Routledge
11 New Fetter Lane, London EC4P 4EE

Transferred to Digital Printing 2004

Simultaneously published in the USA and Canada
by Routledge
29 West 35th Street, New York, NY 10001

© 1999 Lynette Hunter

Typeset in Times by Routledge

British Library Cataloguing in Publication Data
A catalogue record for this book is available from the British Library

Library of Congress Cataloguing in Publication Data
Critiques of Knowing: situated textualities in science, computing and
the arts: Lynette Hunter
Includes bibliographical references and index.
1. Knowledge, Theory of. 2. Science – Philosophy.
3. Artificial intelligence. I. Title.
Q175.32.K45H86   1999
98–30839
121–dc21
CIP

ISBN 0–415–19256–0 (hbk)
ISBN 0–415–19257–9 (pbk)

Contents

Acknowledgements

Books come from collaborations, and this one from work on common ground to build several diverse communities. The thread through all of them has been my partner Peter Lichtenfels, and our two sons Andrew and Alexander, who have, literally, grown up alongside the development of many of these ideas. Building common ground usually means changing what was there before, and it's never easy, but these people have kept me honest and shown a remarkable taste for the uncertainties of exploration. This book is infused with their perception and I thank them with all my heart.

It is quite possible that the book wouldn't have happened at all were it not for Elizabeth Paget who carefully and with great tact helped me over many years prepare the various parts of the text for publication. I am grateful to her, and to Stephen Morton who compiled the index with exceptional skill and an uncanny knack for reading the categories of someone else's mind. Gresham College, where I currently hold the Professorship in Rhetoric, has been generous with financial help for moulding the book into its present shape. I also benefited from the Women's Studies visiting lecture programme at Dalhousie University and a three week University of Leeds exchange fellowship with the universities of Alberta, which allowed me to complete the final draft.

There are so many people in the diverse communities that emerged from writing and reading – the bibliography is the only approximate guide to those whom I should thank. At the same time I would like in particular to thank those who read parts of the manuscript, Margaret Beetham, Bryan Cheyette, Sarah Hutton, Rick Jones, Peter Lichtenfels, Jerry Murphy, Hilary Rose, Stephen Rose, Jan Swearingen, John O. Thompson; and Lesley Johnson with whom I discussed many aspects of the book. I would also like to thank a number of groups: the scientists I worked with including Drs Eidinger, Hall, Rahtz, Smith; Lydia and Bill from McMaster University; the Royal Edinburgh Infirmary; the computer scientists and computer service staff at the universities of Wales, Edinburgh, Oxford and Leeds; the intellectually challenging world of the International Society for the History of Rhetoric; my co-workers on the nineteenth century domestic bibliographies, Dena Atter and Elizabeth Driver, and on the Victorian Periodicals Project; and the inspirational feminist community at the University of Calgary. My friends and colleagues teaching commonwealth and postcolonial literatures at the University of Leeds, under the direction of Shirley Chew, have provided a peculiarly apt location for the situated knowledge of this text and its critique.

Introduction

This book will attempt to offer a study of textuality, particularly in the sciences and artificial intelligence in computing science which has cast itself in the role of a discipline which self-consciously examines the way that science represents knowledge. Textuality allows us to explore the ways we represent aspects of our lives within conventional systems. It also helps us understand how we articulate values that do not have representation. Like knowledge, textuality is bound into power, for textuality is both a mode of knowing and the way we articulate knowledge.

Knowledge and textuality are usually taken as neutral areas by analyses that work within an institutional structure that obscures the connections with ruling power. This is what I call the ideology-subject axis, which I explore in detail in Chapter 1, and an understanding of its rhetoric lies at the centre of my argument and my understanding of the recent development in standpoint theory of a concept of critical realism. Within systems of inquiry that foreground particular connections with ruling power, such as many discourse studies, knowledge and textuality are often taken as determined or constructed or constituted by an ideological system. These studies, implicitly at least, depend on the notion of a constituted subject, and some, with extraordinary insensitivity and blindness have come to define those people outside of the system as 'abject'. However, within the theory of situated knowledge and in standpoint theory, knowledge and textuality are taken as engaged in by groups of people working on negotiating questions of value and action among relations of non-ruling power, and between the non-ruling and the ruling.

In the process of studying the place of textuality within science, I am also concerned to look at what I call the 'gesture to the arts' made by both mainstream and standpoint critiques, and to explore the place of textuality in a number of disciplines. The final chapter of the book extends the standpoint critique in feminist science and technology studies into aesthetics. The insistent gesture to arts strategies of beauty or plurality as a possible resolution for science, does not understand that much recognised 'art' is just as systematic as science, that 'beauty', like the 'success' of science, is also an artefact. But aesthetics is also a field that is tied to the Greek root of the word, to 'feelings', which are generated by all disciplines, and I would agree with Alison Jagger that emotions or feelings are often unauthorised modes of knowing: the 'rational', for example, being an authorised emotion. Aesthetics and epistemology are closely intertwined, for without articulation knowledge remains tacit, and the main focus of the extension of standpoint theory into aesthetics, is to argue for an understanding of 'situated' textuality, analogous to situated knowledge. Situated textualities are where people work on words together to build common ground for the articulation and valuing of knowledge, and to argue for them I draw not only from contemporary social history of science but also from the history of rhetoric.

Standpoint theory argues that knowledge articulated from the standpoint of those excluded from ruling relations of power is particularly important. Because of the exclusion, the knowledge that is offered from that excluded position is quite different to that current within the ethical and ideological systems of a society and its culture, and is therefore a source of assessment and potential change and renewal. The theory is concerned with articulating situated knowledge, with retaining a concept of the real in the sense of critical rather than naive realism, and with re-defining the 'individual' to account for people who are not subjects, or to account for the not-subjected of people's lives. It is important to say, however, that situated textuality and standpoint theory are not special case strategies. All forms of knowledge may be analysed in these terms.

In science, where standpoint has been explored at length, there is in recent debate and as the later chapters discuss, a critique of scientific objectivity based on its self-limitation resulting from the exclusion of, among other things, women's knowledge. In politics, there have been critiques of the curious doubleness of the autonomous yet universalist man constructed by the liberal democratic social contract, because the necessary isolation of that individual obscures the situatedness of their lives. In philosophy, we find the critique of value-free assumptions in both empiricism and idealism, because the notion of 'value-free' denies history. And in the social sciences, there has been the debate between quantitative and qualitative methodology, the latter arguing that verisimilitude, repeatability and enumeration evade the contextual pressures of living.

In each case, the obscured, evaded, denied, excluded situated knowledge is without authority, often if not usually, without words. The critiques delineate tacit knowledge of various kinds, and all recognise the need to work on words to bring those tacit knowledges into communication. In nearly every case the pathway out toward agency that the critiques offer, is through story, narrative or poetics, yet there is no parallel critique of the aesthetic constraints on these materials. The result is a philosophical hiatus that gestures toward the arts but with no concept of the situated textuality needed to articulate situated knowledge, its contexts and its value.

Rhetoric is a field that insists on the bringing together of textualities, society and politics. It has traditionally been concerned with social context, and has always distinguished between the situated, the systematic and the authoritarian. It is also concerned with different approaches to truth: truth as certain, as plausible and as negotiated or probable. And rhetoric is also concerned with the ways individuals and groups wield power, the ways they limit and extend the possibilities of human interaction. Throughout the book I turn to writings on rhetoric from the classical period, and particularly to the texts on rhetoric by Plato and Aristotle. Much of the elaboration of the political structure of democracy is bound up in those rhetorics, as is the development of epistemology and attitudes to value. This infusion of the classical throughout many discourses has led to virtual images of Plato and Aristotle that are formed in each historical place and time, answering to its needs. All of the ensuing discussions return to these classical texts, partly to help dismantle unhelpful screens in earlier metadiscourses, and partly to rebuild alternative versions more appropriate to today. Much of my reading of Plato and Aristotle affirms and lends rhetorical weight to situated knowledges and standpoint theory. In bringing a refreshed tradition to a contemporary need articulated in feminist theory, the rhetorical analysis also moves the theory from epistemology into aesthetics, both in my critique of the aesthetics of different disciplines and in my exploration of different kinds of textuality: textuality as inadequate and hence merely a code, textuality as (in)adequate and hence transgressed and transcended, and textuality as necessarily limited by the materiality of language and therefore the ground for common work on words.

Standpoint theory comes close to discourse studies in many of its concerns, yet the concern for critical realism separates the two. Discourse studies are profoundly caught up in the constitution of social systems, and find it difficult to deal with the notion of the 'real', for reality is messy and requires that systems relax and get snarled up in the nets of living. Yet each does vital work: standpoint theory in its focus on non-ruling relations, and discourse studies in their focus on the constraints of ruling power that constitute individuals. To distinguish between ruling and non-ruling relations, is to distinguish between areas that need different kinds of rhetoric, that manifest themselves not only in different kinds of knowledge but also in different kinds of textuality. And explicit through the book is the political context of the liberal democratic social contract that underwrites the dominant modes of knowledge and textuality with which I live, and which it mediates through the rhetoric of the ideology that represents many western nation states to their subjects.

The first chapter of this book analyses the ideology-subject axis and its rhetoric, in order to assess the connections of knowledge and textuality with ruling power. I argue that ideology is the ethos of the nation state. The strategies of this ideology are common to plausible rhetoric as defined throughout the history of recorded rhetorics. They include: the assumption of common grounds rather than active discussion and agreement to them; the veiling, hiding and obscuring of the constructed status of these common grounds; the isolation of the system to protect grounds from question and change; and the procedure of arguing always within the system, always from the accepted common grounds. When one knows that this set of strategies is happening, the rhetoric can be recognised as an expedient rhetoric that is often successful.

At the centre of the rhetorical stance of ideology is the assumption that representation can be adequate to a lived reality rather than a set of negotiations around the limitations of language. It implies that communication, here political communication between the individual and ruling power, cannot be negotiated differently, only shifted to greater adequacy. The practice of this assumption develops into the concept of the isolated, autonomous subject, increasingly constituted by ideology as the state acquires the stability necessary to its legitimation. Further, as it develops stable representations, it leads to a focus on visual accuracy and repeatability that underwrites the concept of objective knowledge. The ideology of the nation state is common to both politics and science.

A constant thread through all the chapters, is the recognition that much recent political theory takes science as the 'best-case' for politics. And, as I examine in detail in the last two chapters of the book, feminist standpoint theorists in the social studies of science offer direct critiques of the political systems that support it. Science is appealing to analysts of the representative democratic state, since it works with a stable set of parameters that enclose its grounds, isolate its community and allow it to be subject to a rational logic that achieves its success by gaining legitimation from the structures of the system that generates it. Hence issues of legitimation, and how science or political systems justify themselves to their constituents, become central. Modern science has achieved this stable state through increasing involvement in industry and in commerce that need stable technology to maximise profit. Industry and commerce also need stable politics, and since the seventeenth century capitalist nation states have achieved this stability though ideology. Chapter 1 outlines the rhetorical stance of ideology, which stabilises the representation of those in power as well as the identity of individual citizens who become subject to that power. I argue that the stance purposively excludes some social relations and most communities in order to remain stable, and constructs an axis of representation, relating ideology to the subject.

Aristotle describes exactly this phenomenon of communication in his discussion of the rhetoric appropriate to 'science', by which he means the conceptual knowledge developed within a small group of people, rather than an experimental method that could be used by many. He spells out the strategies and devices that are used to construct stability and repeatability, but he also says that such rhetoric is not appropriate to social interaction because it is enclosed, and hence obstructs negotiation and the discussion of differences necessary to political action. Hence expedient rhetoric aimed at success is not appropriate for politics because of its potential for coercion, demagoguery and force. What is interesting is that with the development of ideology in the modern period of western history, politics assumes a rhetorical structure similar to that of Aristotelian 'science', and for the same reason. For three centuries, the politics of western liberal democracies was a club culture, with an extremely small proportion of the population representing a slightly larger number, ruling the whole, inevitably from their own experience and regretfully in their own interests because there were no permitted competing views. Aristotle was not so worried about this because the dominance of the oral medium meant, to him, that the powerful would not stay long in power. But the rhetoric does not work out that way in a period of nation state consolidation with capital, and with the print media and other technologies distancing discussion and stabilising representation.

Central to the argument of this book is that both ideology as representative of the state, and the language that represents scientific practices, use similar rhetorical strategies and devices but for different reasons. The experimentation at the heart of modern science, which has come to define the 'natural sciences', is in effect the primary medium of scientific textuality and should locate its aesthetics. But in conveying the activity of that textuality, science uses language to communicate to other people. However, modern science developed during an historical period when theories of language were attempting, unsuccessfully, to achieve full adequacy to the real, and since language is always different to reality, it appears always to be inadequate. The response of science to this dilemma, which is discussed at length and in detail by scientists during the seventeenth and eighteenth centuries, is to use verbal language as a code, and to employ mathematical language, which of course appears to be more exact because it operates wherever possible within a predefined terrain. Other scientists can understand that there is a real world with which the experiment engages, but for the large part of the population which is unfamiliar with experiment, the second-order code of language is the reality of science.

In Chapter 2, I argue that today, this understanding of science is exacerbated by the recent development of computing science and particularly of artificial intelligence (AI). Neither computing science nor AI deal immediately with the natural world; they are, not surprisingly, often linked with mathematics and philosophy. However, the rhetoric of the textuality of computing mimics that of the second-order code of science and takes language as inadequate to reality. The widespread use of computers throughout the modern nation states of Europe, North America, Australia and Japan, at the least, has given weight and practice to this understanding of science. Hence also, many analyses from political philosophy, in their reference to the scientific model of language and rhetoric, are taking an artefact as a best-case example. They tautologically use an ideological structure to justify the structure of ideology. By way of a rhetorical analysis of the textual representations that science uses, most distinctively demonstrated in computing science and especially in AI, Chapters 2–4 offer a critique of the techniques frequently used for representing what science does. At the same time, using the example of humanities computing, these chapters study the pervasiveness of the belief that these techniques are indeed what science has to offer. The commentary suggests that the persuasion is particularly effective when adopted by the arts and humanities where textuality is not used as second-order code but as primary material, albeit within an aesthetic system which also treats language as (in)adequate and strives always to transgress or transcend it. It has to be said that many people working in the arts and humanities have rejected computing precisely because it mimics the set of epistemological conditions that they challenge, but with hypertext, which seems to offer different strategies, some of that resistance has begun to break down.

The third chapter turns to a study of what mainstream philosophy of science that is concerned with computing and AI, says about the representations used by science. The argument here suggests that these philosophers of science, with a few exceptions, underwrite a notion of textuality that explicitly draws on the (in)adequacy of language and results in representation either as second-order code or as transcendence/transgression. These philosophers also frequently gesture to beauty or plurality as a possible resolution for science, and do not understand that much recognised 'art' is just as systematic as science. However, this misunderstanding is not surprising, since many critics and intellectuals who comment on art do not themselves recognise the systematic structure of 'beauty' nor the ideological construction of plurality. Indeed the one is frequently offered as a resolution to the other, in an essentialist–relativist standoff that is redoubled by political and aesthetic debates about authenticity as identity politics as against multiculturalism.

Chapter 4 works from a rhetorical critique of dominant theories of language and textuality as (in)adequate, to suggest that there are different kinds of first-order textuality or textuality as primary material, and that the different textualities posit different kinds of legitimating practices for knowledge, different ways for people to represent value and to value representation. The critique returns to computing and to a case study from hypertext methodology to emphasise that no technique is enclosing, isolating and reductive, or exploratory, contextualising and flexible, in itself; nor is either authenticity or self-reflexiveness in itself enabling. Communicative texts from all disciplines need a rhetorical analysis of stance, which will position the techniques and strategies historically, politically and socially. Such an analysis situates the textuality, and in doing so situates the knowledge.

The concept of language as (in)adequate to the real is also central to theories of aesthetics and criticism in the post-Renaissance period. Chapters 5 and 6 move through a standpoint critique of science to a critique of aesthetics. Within the modern nation state, the artist is cast as the allowed or permitted transgressor of ideological enclosure, and intellectuals or critics are those who articulate to the state the kind of transgression being enacted. Art itself is perceived to be activity that attempts to transcend the inadequacy of ideological limitation. In doing so it produces beauty as it wrests some element of social reality from ideological obscuring into cultural articulation. At the moment of articulation such work produces intense joy, and partly because it sits so neatly into the interstices of ideology it is often called 'truth'. Yet none of these designations account for the personal and individual work on words and other media carried out by artists. As I argue in Chapter 5, these designations are critical theories of reception and intervention by and for an art that is produced by licensed citizens of the nation state.

What is missing from these critiques, and what this book will move toward, is a critique of the critical and aesthetic discourses for talking about communication, textuality and the arts generally. Without a standpoint critique of the arts, the gesture toward art's strategies made by commentators working in other areas, implicitly takes all those strategies as a good thing, whereas anyone working in the arts knows that some are more appropriate than others. Furthermore, on the whole, the arts themselves are uncritical about the way they present themselves to the public, and without a better understanding of what the arts do in terms of textuality, the understanding of situated knowledge will come to a standstill.

Without an understanding of rhetorical stance and the situatedness of textuality, standpoint can and has been dismissed as identity politics. Yet with a concept of situatedness that rhetoric can offer both to knowledge and text (and to other areas beyond the scope of this book, such as sexuality), personal experience can be positioned with respect to conversation, decision, action and value, within specific groups of people having specific needs. A rhetorical understanding of that situatedness also makes it possible for communities to negotiate with other communities in a larger political field.

Rhetoric argues that language is inexorably different to the real world, which is why language has to be worked on in specific contexts of negotiation over communication or in situated textualities. As a result, I present this analysis as one also based on my own experience as a practising biochemist from 1968–78, as a humanities computing teacher and user from 1980–95, and as a writer and artist all my remembered life. The structure of the book is interspersed with stories, anecdotes and accounts of my personal engagement with these disciplines. The interspersals are at first abrupt as they erupt through the density of memory, and are most fully told in the final chapter. There I speak about my current teaching, in which, with the help of others, I try to educate other people in strategies for conversation and common action with other texts, other knowledges and other people, of quite different needs and positions. The concluding discussion begins to ask what science and computing might be like, what the recognised arts might be like, and indeed what politics might be like, if they engaged more substantially with rhetorical negotiation, probability and community.

1 The ethos of the nation state

Ideology, discourse and standpoint

Over the last thirty to forty years in Europe, one of the main sites for discussion of politics has been the field of the theory of ideology. This discussion will attempt to outline a rather different emphasis, implicit in many of the writings by European political theorists[1] about the relationship between nation-state political theory, the psychoanalytic vocabulary for discussing the subject, and the increasingly rhetorical commentary on ideology. To do so I would suggest that ideology can formally be considered the ethos of the nation state. Indeed now that the totalising definition for ideology so current in the 1970s and 1980s[2] has faded somewhat, the word in its more restricted sense has helpfully become indicative of a specific set of political practices to do with how nation states represent themselves to the people whom they represent, through economic, legal, educational and health institutions, and through the images of commerce, sexuality and social normality that they encourage and disseminate through the media.

The argument of this chapter outlines the rhetorical stance of ideology, which stabilises the representation of those holding power in the state as well as the identity of the citizen who becomes subject to that power. I argue that in order to remain stable, the stance of ideology purposively excludes social relations and communities in areas not to do with ruling power, and constructs an axis of representation relating ideology to the subject as an autonomous individual. Psychoanalytic vocabulary, developed by and on behalf of the citizens of nation states into discourse theory, is highly appropriate to an exploration of this axis. However, neither ideology nor discourse theory attend to the way that individuals work in areas of non-ruling relations.

IDEOLOGY AS ETHOS

Interesting things begin to happen when we look at the implications of taking ideology as ethos. Yet in classical rhetoric, 'ethos' is structured to deal with political communication that is quite different from the dominant political communication of nation states. If contemporary political theorists have focused on issues of legitimation, social and cultural theorists have emphasised the artifice of the nation state. To some extent the argument here reiterates a commonplace of the history of rhetoric: that rhetorical analysis provides an enabling and practical site for interdisciplinary study.[3] This discussion will explore some of the effects of bringing the two fields of politics and socio-cultural theory together in a study of rhetorical strategy, for generalisations about ideology are remarkably similar to commentaries on the strategies, techniques and devices of ethos. However, nation states as they emerge in Western Europe in the post-Renaissance period, and I shall take Britain as a particular example throughout, are constructed in ways that are different in kind from classical rhetorical situations. So to start, I would like briefly to examine some assumptions about political ethos in classical rhetoric, particularly in Aristotle.

Within a rhetorical analysis of any occasion, ethos works like a topos: a common ground on which, or from which, argument and discussion can proceed. There are many ways of establishing a topos, but the key element is to discuss and agree upon what is held in common and what indicates difference. The smaller and more intimate the group, the more can be taken for granted as 'in common' with any negotiation focusing on immediate differences in the social, cultural or political. Aristotle describes for example in *Topica* a relatively small group of people, in privileged positions within the community, in a hierarchy certainly but within a hierarchy of the powerful that explicitly excludes most voices such as those of the slaves, of strangers and of women. Within a relatively small group, if there is disagreement about ethos, then this group can often perceive it quite quickly. They stand physically next to each other; they can nudge each other, disagree and respond in varying degrees depending upon the formality of the occasion. This interaction of course is largely dependent upon the immediacy of the oral context, but similar responsiveness is found today in local media such as community newsletters, yet works in a quite different context, for it has no direct access to political power. The consensual ethos toward which such strategies gesture depends upon members in the group being status-equivalent so that the effect of money and/or power on persuasion can be minimised. The contemporary writer H.-G. Gadamer takes to Aristotle's *Topica* like a duck to water because his political theories concentrate on small communities, and significantly, do not elaborate on the relationship between those communities and the state.

Aristotle speaks extensively about the detail of these interactions within the group that move toward establishing common ground, in both *Topica* and *Rhetorica*. Significantly, he distinguishes between the political and the scientific ethos, saying that for science ethos is not so important because its grounds are taken as self-evident, they are common by prior agreement of the group. More will be said about scientific rhetoric in the following chapter. In contrast, for political discourse the strategies for establishing ethos and other topics are fundamental; they are in effect a procedure for legitimation or justification of political action. The distinction between science and politics is not concerned with the broad legitimation of persuasion itself, but within a particular persuasion, with the legitimation of the position of authority, of author or rhetor which is the ethos of the speaker and with the establishing of ground of general interest, or the ethical field of the group. From the topic of ethos you know a lot about what is to be taken for granted, what may be negotiated and what may be agreed upon. For Aristotle the favoured political ethos moves toward consensus, although he also acknowledges and describes the possibility of scientific or corporate ethos occurring in the political, and the frequency of the authoritative ethos taken by the tyrant.

In the late Renaissance, just at the beginning of politics that combine the nation and the state in Western Europe, legitimation is one of the most pressing questions. As travel and distribution of cultural goods became easier, social interchanges rapidly multiplied and were exacerbated by the growing transfer of power from the church to the increasingly important guilds, city states and courts. Much of the elaboration of rhetoric at the turn of the sixteenth century by writers such as Erasmus, was in response to a growing number of bourgeois citizens and a concomitant diversification of common grounds for argument and agreement. Erasmus distinguishes in a manner similar to Aristotle between consensus, corporate, and authoritative or forceful, ethos. The responsible citizen was fully educated in rhetoric precisely because he had to argue, persuade and agree in the most engaged possible way. *De Copia* can be read as a handbook for pluralism: how to agree consensually about common grounds in a diversely empowered society.[4]

With a larger group of empowered people and the rhetorical potential of the printed medium for writing, ethos becomes simultaneously a site of enormous power and of complete deferral of power into pathos, which is in a sense an abdication of power to the audience. The early modern period records the latter in its unusual production of theories of hermeneutics which focus on the strategies of the reader or interpreter, the audience rather than the rhetor. But since neither nation nor state politics can be legitimated by pathetic agency, the variety of political ethos in the Renaissance moves toward modes of power described by Machiavelli in *The Prince*.[5] Whether read as ironic, parodic, descriptive or prescriptive, *The Prince* outlines an ethos of authority rather than consensus, yet not an authority of overt tyranny and force, rather, a veiled authority that claims to represent the best interests of the state, here most directly the city state. It is interesting that Machiavelli described the rhetorical structure of the power of the monarch precisely at the time it was disintegrating: his analysis could be seen to help the disintegration as well as be a result of the disintegration; in other words the description could not have been written before the concept of political legitimation residing in the ethos of the monarch had begun to collapse, and yet because of that collapse Machiavelli is able to articulate the strategies that veil the monarchy, give it an apparent coherence and stability.

From the seventeenth century onward, larger and larger numbers of people laid a claim to political empowerment. It is possible to follow the institutionalisation of the public discourse of politics in terms of a combination of nationalism, which had been present from much earlier times, and capitalism, that between them effectively control that diversity. Capitalism needs a broad power base to encourage competitiveness between players who have roughly equal opportunities.[6] One can read capitalism as an extension of the smaller court system or the city state into national politics:[7] bringing to economics the broader franchise of bourgeois political structures. The envisaged commonwealth was no small dream, but its procedures for self-moderation soon became an end in themselves.[8] The ethical position of capitalism is to allow for diverse voices within a political system that can deal with potential conflict between competitive voices, or at the least to translate the viciousness of acquisitive contest by force, into a rabid exploitation of economics.[9] Capitalism benefits from nationalism not only because the latter provides a rationale for such a system but also, specifically, nationalism offers a protective device for some markets, and is successful not only at stabilising potential conflict among the empowered but also at stabilising public and popular demand. Later on, when the exploitation that is the source of capitalism's profit comes closer to the nation, the techniques for stabilising the public enable a self-commodification which is the central contradiction for contemporary capitalism. This analysis is not new, but we need to look at its implications for rhetoric.[10]

Nationalism combined with capitalism into a politics of the state, is a specifically modern event, and has a number of problems with rhetorical ethos. First, there is the issue of distributed power: the group of people in power are not from a small coherent part of the social hierarchy; there is more than one of them and there are many more people to whom they have to answer, and there will be diverse and at times contradictory points of view which all claim representation.[11] Second, the political ethos cannot be directly responsive to its audience either through an individual in an oration, or in a writing. The ethos cannot be consensual; its conditions make it necessarily corporate, counting on prior agreement for decision, and action taken on behalf of others. The various parliamentary systems in Europe have emerged specifically to cope with these conditions, and the notion of representative government in particular raises rhetorical problems rarely addressed by contemporary political theorists until the recent discussions about individual activism and community agency.[12] Given the conditions of this ethos, there has first to be a way of stabilising the reception of the representation of the group in power, giving it its veil of authority, and second, a way of stabilising the description or definition of the individual within the nation. These are the political implications of the rhetoric which match the needs of capitalism to stabilise or contain the conflictual powerful and to stabilise and maintain popular demand. In a sense, the rhetoric provides not only an ethos for establishing the presented 'character' of the rhetor, but also guidelines for establishing a specific responsive state in the audience, or an ethics of pathos. Its ethos is a strategy and imposes a strategy: a particular stance that lies within the techniques and strategies of fantasy which denies the possibility of any intrusion from reality external to its grounds. What follows attempts to indicate how this can be effected.

STABILISING THE GROUP IN POWER: IDEOLOGY

It is interesting to note the urgency that from the seventeenth century marks the need to find stable representations. These are most often studied in science and philosophy under the banner of Descartes' description of philosophy *as* representation, or the attempts by the Port-Royal logicians or Royal Society scientists to develop systems for accurate presentation of the actual. This attempt becomes a central focus for my discussions in the next two chapters of the rhetoric of computing science. Yet I would argue that simultaneously the political and then increasingly the institutional structures of nations in Western Europe, albeit at different times, move toward state systems because they too enable a means of delivering stability.[13] State systems as they have emerged in Western Europe offer ways of constructing the corporate ethos that is being explored here as 'ideology', and that ideology has quite specific rhetorical strategies and structures: primarily, its stability of representation depends upon the public's willingness to forget that it is a representation. This is achieved through a claim both to neutrality and/or truth, as well as to natural, transparent presentation of the real.[14] With only a slight reversal of Machiavelli's emphasis, the veiled illusion of absolute power held by the monarch becomes the representation of stability needed by nationalism[15] and capitalism within state politics.[16]

Documentation and exploration by cultural and social historians, often by way of Michel Foucault's work on post-Renaissance systems of surveillance and punishment,[17] is a recognition of the structures that demand stability and emphasise norms within the voice of institutional public discourse. Foucault also suggests that the stable institutional representation of the state which consolidated during the eighteenth century, was fundamentally challenged by the liberalism of say, an Adam Smith, that insisted on the limits of the state. However, the articulation of liberalism in the late eighteenth century was effectively an explicit statement about the implicit interrelationship between the nation state and capitalism.[18] Its explanation of government as an economic system was predicated on the particular combination of state and capital.[19] At the same time, Smith's recognition that the regulation of capital and the regulation of the state were different even though bound symbiotically together, was important for a different reason, because it opened up the apparently natural representations of the nation state to an understanding of their artifice, and hence to a systematic exploitation of their power to stabilise. It may be important that Smith was a rhetorically trained political economist, living in a Scotland which had recently been annexed to England. In other words he would have been well positioned by his education and his political history to recognise the artifice of nation-state ideology.[20]

Control over the stabilisation of representations of the powerful became particularly pressing during the early nineteenth century, with the capitalist need for stable demand in order to make technology profitable at a time of expansion in capital-intensive heavy industry. Anthony Giddens has argued that it is in effect the combination of capitalism and liberalism with industry that brings about the exploitation of the working class by way of the political system. The argument here will work, as others have, from this insight to other systemic exploitations such as those of women. But the rhetorical point is different. The strategies adopted by the capitalist nation state, before and after it becomes thoroughly integrated with industry and business, are the same. The difference lies in the extensiveness and impact that those strategies, which I am calling ideology, have upon the people of the nation. Giddens goes on to speak of class relations as contractual relationships, and it can be suggested that these relationships are part of the formation of subject positions.[21] It is no surprise that this system of commodification of desires is articulated by Marx just as those demands begin to exert widening pressures. Of course ideology affects not only those involved in contractual relationships such as working-class men, but implicitly embodies substantial numbers of people excluded from participation within the state, in the shadows of its visible formations.[22]

If much of the work by political theorists has focused on issues of legitimation of power and authority within the nation state, social and cultural studies have emphasised its artifice. What is being attempted here is a way of bringing the two together to bear more closely on each other, asking what the implications of those legitimating strategies are for effective political rhetoric. Simply put, the way ideology solves the legitimacy problem of the government of large areas and diversified voices, is to deny/evade/ignore the need for responsive ethos. How this comes about is various, but a rhetorical analysis of the ways it does so is essential. A congruent analysis marked by its clarity and relentless exposure of the taken-for-granted, can be found in Carol Pateman's critique of the liberal democratic social contract. She argues that there is a two-stage process in the structuring of liberal democracy during the seventeenth and eighteenth centuries in England: the first comes about when free and equal individuals come together in a political community and agree to form a common political authority; the second moves on from this agreement when the individuals give up their common authority to a few representatives (1985: 52). The former is a consensual agreement, while the latter is corporate.

What Pateman goes on to discuss is the way that for Hobbes the two stages are difficult to distinguish, because the stage of consent is part of the submission to the authority of the state which is necessary to the survival of the individual in the marketplace. Furthermore, for Hobbes, the state or Leviathan, becomes one with the people because men are forming themselves into a unified body under contract and civil law. What is interesting about this concept is that it offers at one and the same time the idea of absolute representation of the individual in the state, and the idea that to form a unified body by accepting the state as representative of all, it 'makes the person one', it divests the person of multiplicity and renders it subject to representation (68). The result of the elision of the two-stage process is that people (men) become free and equal at the same time as they also become subject and incur tacit obligation to the state (52). Hence legitimation is not an issue. Pateman's argument goes on to elaborate by way of Locke's distinction between genuine consent and tacit consent, the division of the public world into political government, where representatives are chosen by the propertied who thereby consent 'genuinely', and the civic which is made up of people who lack enough property to vote, and hence only consent 'tacitly' to state rule (71).

It is doubly interesting that Pateman's critique exposes the tautological structure of Hobbes' logic, as he argues by analogy with geometry that since the structure of the commonwealth is also man-made it is in the hands of individual agency distinct from other social forces. He 'argues from a result to prove it' (39). Furthermore, Locke's introduction of analysis from economics is entirely concerned to justify assent to government whose legitimacy is already assumed (62). The common grounds hidden by Locke are found explicitly in Hobbes: the reformulation of patriarchal power from paternal to fraternal by way of domestic and sexual contest (65). Men and women are initially equal, but men acquire individual agency by becoming masters in their families; the conquest of conjugal right is at the heart of civil society (68). From the start, the liberal democratic state separates the public, with political and civic rules and rights, from the private or domestic. Those giving tacit consent to the state in the civic, and those excluded completely from it, become carriers of tacit knowledge within the ideology of the nation state, that cannot be voiced because they are not citizens of it.

Extending from the work of Carol Pateman and many others particularly concerned with the racial and ethnic grounds for the concept of 'nation', Nira Yuval-Davis points out that the state is not unitary, even though, when the state can persuade people that it is coincident with 'nation', it can appear united and 'natural'.[23]

David Beetham comments that any analysis of legitimacy within capitalism that concludes that it is achieved by consensus, utterly fails to look at how that construction is reached. Beetham moves on to suggest that it is not consensus at all, but a corporate control of political power that represents itself to the public as consensus.[24] If the establishing of common grounds is in part how words acquire meaning, it becomes the primary way that a government establishes its legitimacy. But even when common grounds are agreed to, conflicts in meaning can emerge, and consensual agreement encourages a greater awareness of the potential for this conflict and possibly even a greater engagement with it than the corporate. Hence consensual agreement is far less effective a way to attain stable legitimacy than the corporate, yet the model for a political ethos stresses the consensual as the most valid. Indeed, even the word 'ethical' connotes active, participatory agreement to a working common ground.

However, within a representative government, consensus is not only impracticable but downright unhelpful, so the governing body must either grant this and deal with the outcome, or hide it. A corporate system that acknowledges that it is requiring normative behaviour may be more likely to respond to and engage with 'normal' individuals, than one that implicitly coerces its members into the conventions of the system and does nothing to follow up on the requirement. Swedish state corporatism might be an example of the former,[25] and Britain or the United States examples of the latter. But neither corporate strategy is necessarily more or less able to engage the individual; both are historically positioned. Beetham distinguishes between communist states which defined their representation overtly and as such, yet claimed a monopoly over it, and capitalist states which initially established their moral authority through legality, justifiability and demonstration of common grounds, yet developed away from these strategies as actions responding to material conditions, toward using them as *a priori* principles, effectively as *status causae*.[26] Tom Nairn suggests that this was particularly so in England because that is where capitalism slowly developed, rather than being taken on as a theory.[27] At the heart of the capitalist nation state is an ideology that claims self-conscious questioning of principles, yet also brooks no disagreement with its truths:[28] an ideology that is fully rhetorical and persuasive, yet which claims absolute stasis, neutrality and authority:[29] an ideology that represents, but then claims the natural and wants to forget the artificial.[30]

Ideology uses the medium of state institutions to imply that there is a norm, a conventional, even a natural state. Because this is presented *as the case*, there is no need to question or interact with the ethos. This is the equivalent to a denial of the need for rhetoric, and, although there are other reasons to which I will return, the gradual loss of formal rhetoric in Western Europe since the seventeenth century, which has been particularly marked in England, is part of the stabilising effect of this ethos. The rhetoric of the capitalist nation state structures its ethos simultaneously to build an artificial norm, *and* to forget that it is artificial. Technically the rhetoric works from accepted common grounds, corporately agreed rather than consensually achieved, and uses representative media that try to repeat them without variation, which action becomes more possible as technologies for stable repetition, such as film, are developed. As with the factory component work of Fordism which introduced an effective standardisation of work through continual repetition of identical actions by the workers,[31] the repetition of the same helps to commodify desire and retard change.[32] And finally, its power is mediated not direct. Hence the source of power can never precisely be located, it is always veiled and hidden.[33]

Machiavelli describes the way that the Prince effects power never directly but always through others, as an example on a smaller scale of this rhetoric. When such dissemination of power can be carried by sophisticated technologies for invariable repetition, the functionalities of 'hiding' and 'naturalisation' are lost; they can become an amnesia.[34] *Mein Kampf* describes exactly this wielding of power through others, and they act on your behalf because interpretation is impossible from authority that claims absolute truth. The corporate becomes authoritative.[35] *Mein Kampf* also describes a limit case where it has to be asked whether *all* individuals within the state are amnesiac, or whether they use the amnesia itself as a legitimating artifice.

STABILISING THE 'INDIVIDUAL': THE SUBJECT

The techniques of corporate grounds and invariable repetition that construct commodification, define part of ideological ethos. Ideology works by building representations that seem to be naturalised ways of life; and within the ethos of the nation state, pathos is controlled by defining the audience as fixed; the individual has to accept that 'natural' state. Individuals do so by remembering to forget the construction of the system and by becoming private.[36] Both as workers and consumers they are built into isolation from other individuals, they are commodified. From the individual's point of view, whether as a worker or as a consumer, this commodification results in a stabilisation of identity. The process has been documented by way of recent psychoanalytical theories of the subject, elaborated on behalf of the subject within the nation by for example Franz Fanon, and on behalf of the gendered subject in many areas including that of Lacanian feminism. Psychoanalysis articulates the problems arising from the construction of a fixed, stable subject, who is the private individual defined as isolated, without community and not immediately responsive. The circumscribed family within a contained physical space is an important device for the successful construction of such a subject.[37] But the key to the implementation of ideology's representations is skill in remembering to forget, what George Orwell called 'doublethink' in *Nineteen Eighty-Four*.[38]

In other words there is no conspiracy, but the combination of capitalism and nationalism finds state institutional structure successful at stabilising representation. Where the ethos becomes questionable is in its attempts to present the artifice as an adequate representation, and hence natural. Pocock offers the example of the way the constitution in Britain becomes a set of accepted 'corporate' grounds by way of Edmund Burke's notion of the state as a family or 'trust'.[39] He distinguishes between Hobbes' notion of the 'natural' state and Burke's description of 'artifice' in the construction of tradition that you may need to invent but then forget the invention. Indeed, it is not until the early nineteenth century that this ethos comes to be recognised as a potentially self-conscious device. Walter Scott's historical fictionalising of Scotland, or Walter Bagehot's construction of Victoria's monarchy are examples of this recognition;[40] and it has been argued that as capitalism's control over representations of the state becomes clearer at the end of the nineteenth century, the intellectual response was to converge upon anthropology as a record of 'true' ritual that could offer an alternative to the artifice of politics.[41]

These specific manifestations of ethos in eighteenth- and nineteenth-century Britain are less monumental than those that accrete around the theories of ideology offered by political theorists of the 1970s and 1980s, but they describe an insistent rhetoric of denial and contradiction infusing the whole social fabric. Perhaps Diane Macdonell's reading of the potentially monumental Althusser is more subtle since she moves along the ideology-subject axis to analyse nodes of discourse:[42] that ideologies come from conflicts but also reimpose those conflicts and are hence the site of a double struggle: that wage earners are obliged to sell their labour power to those who make profit from that labour so that that profit can be made. They support their exploiters; they support the exploitation of themselves as they remember to forget it.[43] Such a class analysis is more fully familiar to cultural and social studies in the parallel questions of gender: how do women accept and support, even desire, their own exploitation?[44] This internal contradiction remains the central focus for much of cultural studies, yet it is rarely analysed as the direct result of rhetorical strategy necessary to the dominant form of twentieth-century politics.

It is no surprise that capitalist strategies for gender commodification in and through advertising, which are structured on directly analogous techniques of naturalising and neutrality, also came into full play during the nineteenth century. Yet as ideology becomes more self-consciously recognised as artifice, the demands that it puts on people to remember and simultaneously to forget are immensely stressful. The strategy of remembering to forget lies behind genres such as the gothic, behind the split personalities of nineteenth-century poetry and prose, and beyond the increasing separation of the private from the public. The strategy also codifies into nationalism the need to locate elsewhere what is forgotten; it gives added status to gender discrimination, it naturalises racism.

It could be said that over the last century, probably due to the enfranchisement, much commentary has been implicitly directed at trying to articulate the strategy of this ethos with its combination of corporate agreement, and commodified stable representations that discourage questioning and ask us to forget. The commentaries also attempt to deal with the fall-out: Engels' elaboration of 'false consciousness' lies at the heart of the contradiction/amnesia; Wittgenstein's careful analysis of the impossibility of stating anything as 'the case' is a direct critique of ideology's claim to represent without artifice; and Freud offered observations on the results of this strategy in repression[45] – whether it be narcissism (total repression), neurosis (the ambivalence of memory) or psychosis (fully fledged externalisation of the forgotten). Central to ideology's success is the concept that language or representation can be adequate to the reality of an individual, that a person can be fully represented in terms of the subject position constituted by ideology. Freud offered an especially helpful vocabulary for exploring the tensions that result from inadequacy, and by implication opened up the possibility that the ideology-subject axis is not the only way of talking about or knowing our lives.

Cartesian dualism, conventionally read, splits the brain/body from the mind. Descartes proposes this as one way of explaining the limitations of language in its attempt to re-present the phenomenological actuality of the world. While Descartes recognises this as unstable, the suggestion gains actual currency as noted above via for example Port-Royal logicians, and even the Royal Society. The possibility of progress toward a stable representation of referential actuality or phenomena, occurs concurrent with and no doubt as part of the political necessity for the emerging nation states of Europe to present a coherent argumentative ethos to each other. It has withstood the tensions of the rational via Kant's relational twist to ideological representations, which throws into relief the elements of structure that need to be readjusted to keep the status quo static. The Cartesian split becomes an appropriate common ground, self-evident fact, and generates two interlinked spin-offs about the body and about language that in contemporary cultural theory are both linked to desire and have specific implications for rhetorical strategy.

If Descartes dreamed of a language that could fully re-present phenomena, he knew the limitations of language as a condition. But those working in Cartesian space took the dream as a possibility. This possibility held within it a multitude of utopias, including the hope that the new political systems of increasing state government could create a large and cohesive commonwealth to replace the feudal. However, while the poets worked differently, the politicians, scientists, philosophers and theologians (who had always had this tendency to forget) began, at least on the printed pages we still keep, to forget that representation is *necessarily* limited, and to think of the limitation of language as a problematic inadequacy. What Freud made accessible to everyone was a vocabulary for discussing this idea that representation is always inadequate.

Since Cartesian space splits the body from the mind, thought becomes representable and separate from the body. Further, the mind cannot deal with the possibility of unrepresentable knowledge: The history of psychology since the seventeenth century is an attempt to deal with the detritus flung from this severance which makes the incommunicable (i.e. not systematic/ideological) 'mad', located in the body, deranged. Just as there are tensions in Freud's simultaneous and contradictory linkage and separation of the body/mind, so there are tensions in his contradictory separation and hope for linkage between the represented and the unrepresented. At the centre of Freud's definition of psychosis, neurosis and narcissism, is the 'other' as unrepresentable. The 'other' becomes allied with the incommunicable body, hence what *is* represented is never complete, never adequate to reality because it does not deal with all phenomena. To represent, to *think*, in Cartesian space is to enter a system increasingly dominated by the need for stable representations of ideology.[46] The more pervasive this ideological discourse becomes, the more stable the representation, the more inadequate the incommunicable. This is the strategy that invents desire, and introduces notions of completion, fullness and absolute presence.

But there are different approaches to the limitations of language made possible by reading Freud in alternative ways. To return to Descartes: by remembering the instability of language, part of the common ground for Descartes' thinking is that 'thought' is a way of working toward articulating the not-yet-articulated. Like 'theory', which in many contemporary discourses tends to get separated from practice but which is in effect the same thing, his 'thought' is trying for appropriate representations of practice: but why? Articulating practice can be understood variously as a way of contacting 'reality' and as a way of making individual practice social.[47] Freud described the incommunicable as the repressed, focusing on two different kinds of repression, into the unconscious (not possible to represent) and the subconscious (possible to represent). The 'unconscious' becomes a concept that describes a response to the sense of a systematic stable ideology inaugurated by nation-state governments that emerge in post-Renaissance Europe. It is a way of providing an origin or raison d'être for the 'private', and links the private with the body, particularly the body we cannot communicate.[48]

If the unconscious is understood as a constructed political response to authoritarian politics,[49] then there is a clear transition into a wide variety of social repressions under state governments which institutionalise community functions. If the state is authoritarian but also powerful and systematic, then the disempowered are not just partially repressed but completely repressed, eradicated from participation. The terms of the ideology-subject axis become: unconscious vs. system; private vs. state; isolated individual vs. nation. But what this also does, apart from providing a political rather than a biological reason for the unconscious, is ally the impossibility of communication with disempowerment.

The alliance has a curious effect on people who are in effect empowered and should thereby be able to communicate because the system works for them. They hold the position they are in because a state system defines them as powerful in a particular way, so if people are empowered then their inability to communicate must be a result of the system. This seems to make sense because the system is presented as a symbolic mode of representation that is taken as necessarily (and hopelessly) inadequate. It is only those who are disempowered who understand that the (in)communicable is work, and only these know hope. The unconscious as the split self, is the response of the powerful/empowered to their own sense of the hopeless inadequacy of the public representation of the symbolic to their individual and 'private' lives. And it is those who have been *relatively* empowered who have used the 'unconscious' as an analytical tool for criticising systems of authoritative power: for example Franz Fanon or Homi Bhabha on the colonial subject, or Juliet Mitchell, Luce Irigaray and Jacqueline Rose on the repressed woman.

It is vitally important that authoritative systems of power are analysed critically from within their own terms: They cannot 'hear' anything else because it is repressed, absent, dismembered by forgetfulness. But those relatively empowered speakers are in a highly ambivalent position, dependent on the degree of their disempowered status. Within this powerful system of discourse, in order to talk about the disempowered, they have to talk about the unconscious, but in talking about the unconscious they accept the framework of authoritative state vs. private that creates disempowerment. If their aim is to interpellate a disempowered subject into the representative system (already taken to be inadequate) this is valuable because then that subject can be 'heard', but simultaneously that subject is dismissable as inadequate. This inadequacy only diminishes as the representation moves closer to the sufficiently adequate and the subject is systematised. Yuval-Davis notes that such 'Cultural homogeneity [from a Foucauldian perspective] would be a result of hegemonisation, and it would always be limited and more noticeable in the centre rather than in the social margins' (1997: 41). It is striking that the further one moves from state representation, just as from cultural hegemony, the less helpful becomes the Freudian vocabulary. The determinism of the process has led feminist theorists increasingly toward analyses of how the 'subject' can be changed, not just brought into being.[50]

In the eyes and ears and mouths of the powerful, the disempowered are dismembered, part of the unconscious, the body, the private. Like the unconscious/body/private, they are part of the 'natural', the 'intuitive', the 'primitive', the 'not-civilised', the incommunicable. For example, until recently the metaphors for women make no separation between gender and sexuality because the body defines their position outside the symbolic. Just so, the body defines the position of visibly 'different' people outside the system. This is one reason why class analysis was effective for so long: it was difficult to locate the 'poor' outside the system on body terms, so it was done in terms of the 'private' – although the spurious connecting of a working class with sexual 'perversion' or the cultural battle over fashion are indications of the way the media are used to transfer 'poor' into bodily 'difference'.

The relegation to the unconscious by state ideology, of women and other physically 'different' groups of people had been so effective a political strategy for stable representation that Freud's popularisation of an emerging vocabulary for discussing this repression was of course profoundly unsettling. By definition the unconscious should have remained incommunicable. With psychological and psychoanalytic methodology, a discourse was formed both to enable people to talk about the language as inadequate and why representations are to be distrusted, and even ignored, as in the forgetfulness of ethnocentricity. What Freud pointed to on the ideology-subject axis was the terrible stress that the internal contradictions of capitalist nation states placed on its citizens by representing them, and that those excluded from the axis had no representation at all.

DIFFERENT KINDS OF THE FORGOTTEN, OR DIFFERENT KINDS OF TACIT KNOWLEDGE: INDIVIDUALS WORKING WITHIN NON-RULING RELATIONS

Because the structure of the liberal democratic nation state places its citizens automatically under 'obligation' which is self-justifying and therefore potentially coercive, it is likely that individuals will find any one subject position inadequate to representation. To function effectively in public, within relations of ruling power, each person will carry with them a greater or lesser need to remember to forget things that they know, a greater or lesser extent of tacit knowledge. But there are different kinds of tacit knowledge that result from the different rhetorical stances within which individuals find themselves working. For example, in science, members of a scientific community which is based on a set of common grounds tacitly assume those grounds, tacitly assent to their representative adequacy. Yet however difficult to do, the grounds are open to exploration and foregrounding if they acquire more adequate representation. Similarly, in liberal representative democracies, those people whose interests are most clearly aligned with the capitalist nation, such as the affluent, the white, the male, tacitly assume the grounds necessary to sustain that state, tacitly assent to their representative adequacy. Yet again, however difficult, those grounds are open to exploration and foregrounding, usually by intellectuals or artists, who are recognised *as such* by their subject positions within ideology and are inevitably citizens with some form of adequate representation.

Indeed, a primary function of art is to negotiate the appropriateness of a common ground by way of mimetic activity in whose performance the value is recalled. Art can recall the tacit knowledge that individuals have learned, into a social medium which allows for overt communal recognition and assessment of common ground. Yet the tacit that is recalled has a different status if it lies within or without the ideology-subject axis. If art repeats and refreshes a common ground that has become an assumption, it can save values from the negative definition of common sense, the habitual things that we do not think about. Artists are also the licensed transgressors of liberal democratic nations, cultural work being to wrest into representation those elements that ideology wants forgotten. In doing so, they re-negotiate agreements about the (in)adequacy of representation to the previously tacit, forgotten-because-hidden elements. Yet when states repeat agreements by institutionalising their performance, the ethos again becomes very stable, frequently it seems discouraging of the private memory of performance and transgression. Private memory may become subversive in the face of the public memory of history.[51]

At the same time the ideology-subject axis also completely excludes many individuals from any form of representation, and from the position of subjects in the axis those voices simply cannot be heard. They form an area of knowledge that is impossible for intellectuals and artists to communicate. Yet from the position of those excluded from representation, whether adequate or inadequate, who are working alongside the relations between individuals and ruling power in areas of non-ruling power, communication of their lives is certainly not impossible. Neither is the communication to do with representation, but with working on the limitations of language and words to hammer out agreements among individuals, not just from one private individual, to appropriate articulations. In effect, work on common ground that is tacit can be carried out for a number of reasons. Either because it lies beyond ideological representation, or because ideology wishes to forget it or wants to assume it. But trying to communicate the tacit is the only way to begin to differentiate among the different kinds of representation and articulation. I will discuss the difference between these two stances in Chapter 5. It is not surprising that personal memory is so closely tied to artistic work.[52]

However, responses to different kinds of tacit knowledge have been far more defined than those to different kinds of textuality in art, despite the way that Freud's 'unconscious' has been used to explore the repressions of domestic space, and to encourage people to attempt articulations of the 'different', often using the body as the site for alternative articulations, as in feminism's 'writing the body': re-membering and dismembering the individual in Cartesian space. If Freud provided in psychoanalysis a vocabulary for analysing how relationships of power and knowledge are negotiated between the individual as a subject and nation-state ideology, Foucault has been equally influential in laying out a vocabulary for understanding those relationships in terms of how subjects come to be formed or positioned by the state, particularly as fixed and 'the same'.[53] And whereas Foucault is frequently criticised for ignoring the conditions of the private in favour of those of the state, Freud's emphasis on the family at the expense of public politics somewhat redresses the balance.

Not enough distinction is made between Freud's elaboration of a vocabulary that is specific to the relationship between the nation state and its subjects and Foucault's opening out of the concept of discourse. The particular rhetorical structure of Freud's ideology-subject axis is quite different from the much wider field of subordinate discourses which work with and alongside capitalist nation states, but which contest the determinism of that axis.[54] Judith Butler's *Bodies that Matter*, which offers an index to work that has recently been done on these discourses, pursues the distinction in terms of artifice as opposed to performativity, where the latter is a constrained repetition of norms whose very reiteration implies that the individual has not been represented.[55] For Butler, psychosis is a threat of fixity, of artifice: within the terms of ideology's rhetoric, a threat of permanent and destructive amnesiac contradiction. Yet again, if analyses of ruling state power rarely consider the individual's response, social and cultural studies rarely engage with structures of the capitalist nation state, the ethos of which necessitates many of the contradictions upon which they focus.

Like Machiavelli's *The Prince*, the increasing ability to articulate the rhetorical structure and implications of the nation-state ethos, comes at a time when that state is losing legitimacy. The analyses both contribute to and are a result of that loss of power. Concurrently the nation state is losing power to transnational corporations and global economics, and like the monarchy of the sixteenth to seventeenth centuries, the nation state is beginning to be perceived as a cultural artefact, and in some places has even given way to the concept of nation and ethnicity. The political disarray in Western Europe is partly due to the perception that nation-state politics is losing its economic power, so one cannot have a realignment of economic power in terms of corporate negotiation at national level, let alone consensual discussion. Political theorists are concerned largely with the effects of the loss of the legitimacy of the nation state as a result of the transfer of economic power, and with the difficulty of handling the internal contradictions of capitalist democracy when the nation state is subject to global economic forces.[56]

It is perhaps in recognition of this, that social and cultural theorists are moving inexorably into the area of discourse and leaving ideology behind. Yet there have been recent and important developments that shift the focus of 'discourse' from the ideology-subject axis to the individual and non-state power. For example, the standpoint theorist Dorothy Smith reiterates that 'Capitalism creates extralocal, impersonal, universalised forms of action'[57] so that, in a restatement of the need to remember to forget,

The abstracted conceptual mode of ruling [ideology] exists in and depends upon a world known immediately and directly in the bodily mode. The suppression of that mode of being as a focus, as thematic, depends upon a social organisation that produces the conditions of its suppression.

(Smith 1987: 81)

She then goes on to elaborate the position of standpoint theory as working precisely from the local conditions of power that are excluded from state-subject relations: in the civic and domestic rather than ideology. A considerable amount of this work has effectively been carried out by feminist analyses of science and the scientific community, to which I shall return in Chapter 5.[58] As Aristotle noted, science has a corporate ethos; it offers a clear case study for several of the elements in the rhetoric of ideology.[59] While there have been few studies of the relationship between science and the nation state which have looked at both ideology and discourse, the work done on standpoint theory in terms of science is located precisely at the intersection of the individual as subject and as agent.

Standpoint theory is concerned with situated knowledge as necessarily partial, and with retaining a concept of the real in terms of critical realism rather that naive realism – a distinction made most discretely by Hilary Rose.[60] It is also concerned with redefining the 'individual' to account for people who are not subjects, or to account for the not-subjected of people's lives. These emphases lie close to those of discourse theory in two of the three areas, and it is important to distinguish between discourse theory and standpoint theory. Situated knowledges are similar to discourse systems, and both have been accused of relativist pluralism. The notion of varieties of individuals or of variable identity is similar to the study within different discourse studies of the relation of the subject to ideological representation. But the concern with the 'real' separates the two, for discourse studies is profoundly caught into construction and even the constitution of subjects along the ideology-subject axis. Because discourse studies are systematic they cannot easily deal with the notion of a 'real' that is external to the system. And reality is messy, it asks that systems overlap and get mixed up rather than remain distinct.

To understand the grounds of distinction between discourse studies and standpoint theory, it is helpful to bring together the work of Carol Pateman and Dorothy Smith. In 1973, Pateman's profoundly influential critique of the liberal democratic social contract that underwrites many of the representative democracies in the Euro-American west, illuminated the work of many other political thinkers by defining the basis for the liberal social contract as the isolated autonomous individual (also known as universal man), the notion of objective knowledge, and the idea of neutral and therefore unquestionable truth. In many ways the analysis is similar to that made by Habermas, yet Pateman increasingly focused on women and defined the areas of the relations of power as government, the civic and the domestic. Whereas for Habermas the split between the government and the civic interacts in various ways with the split between the public and the private, for Pateman there is the government, the civic and the domestic, interacting with the ruling public, the civic public and the non-ruling public. I understand her concept of the ruling public to include the private, and her concept of the non-ruling public to include the personal.

Whereas for Habermas there are liberty, equality, fraternity and special rights for women and others, for Pateman liberty, equality, fraternity *are* special rights for men. This distinction makes a fundamental difference to any discussion of 'rights' occurring for example with respect to the welfare or social security state. For a long time the liberty, equality and fraternity of the liberal democratic social contract only refers to 5–20 per cent of the population, depending on the extent of the franchise. These are the citizens of the nation, and hence subject to the state through ideological representations which define access to ruling power. Even with enfranchisement, the remaining 80 per cent have only representative access to ruling power, only actual access, if they acquiesce to the permitted representations or obligations which have already been put into place by people not necessarily of their class, colour or gender.

In Pateman's more recent work, and throughout the work of Dorothy Smith, one finds not a critique of ruling relations but a turn to the 80 per cent. Smith points out that most of those supposedly with the 'vote' have no actual access to ruling power. She asks how and why we can value the social and political, non-ruling relations of power that involve so many people; and claims implicitly that that valuing will, through its articulacy, encourage if not allow communication with and possible change of ruling relations of power. For Smith, in terms of understanding the socio-political world of the liberal state, there is ruling power and non-ruling power: she does not elaborate on the civic.

From Pateman and Smith one can derive three areas of the elaboration of power. There are ruling government relations mediated by ideology through representations of the subject – what I have earlier discussed as the ideology-subject axis. There are also the tensely contradictory, distinct systems of discourse where adequately represented subjects (the 5–20 per cent) contest those representations on the edges of government and in the civic, especially in fields related to capitalism. Those representations are fundamentally connected to the representations of psychoanalysis. And there are the non-ruling, civic and domestic relations of power, simultaneously negotiated among and between individuals and groups, which are inflected by subjecthood, however inadequately represented, by the contradictions of systematisation in discourses, but also by local daily communications, discussions and negotiations. Many feminists interested in modulating ideological representations by way of Foucault have translated Smith's 'non-ruling' into the discourse study area that analyses specific systematic moments along the ideology-subject axis, in terms of subjects dealing with their contradictions. While this theoretical move is important, indeed vital, since discourse studies are currently the main voices to which ideology responds, it misses the point that people without access to ruling power cannot be represented subjects. It can also be class-blind. In neglecting those people without subjecthood, it misses the strategies that could be employed as economic power moves from national to multinational to global. This is potentially disastrous, since that move renders increasingly ineffective the rhetoric of discourse studies which is bound to nation-state capitalism.

The three areas delineated by Pateman and Smith, government/civic/domestic, or ruling/discourse/non-ruling, establish a pattern that has been taken up by standpoint theorists who do focus on the non-ruling and excluded. One finds not only the feminist critique of objectivity in science, with its focus on the exclusion of women's knowledge, but also the feminist critique of politics that analyses the curious simultaneity of the autonomous yet universalist man whose isolation obscures their situatedness. In philosophy, one finds the critique of value-free assumptions in both empiricism and idealism, because the notion of 'value-free' denies history. And in sociology there has been the debate between quantitative and qualitative methodology which charges the enumeration, verisimiltude and repeatability (as distinct from the broader and various strategies of repetition) of the latter with an evasion of the contextual.[61] In each case the obscured, evaded, denied, excluded situated knowledge is without authority, and often, if not usually, without words. The critiques delineate tacit knowledges of various kinds, and all recognise the need to work on words to bring those tacit knowledges into communication.

What is missing from these critiques, and what this book will move toward, is a critique of the critical and aesthetic discourses for talking about communication, textuality and the arts generally. Without a standpoint critique of the arts, the gesture toward art's strategies made by commentators working in other areas, implicitly takes all those strategies as a good thing, whereas anyone working in the arts knows that some are more appropriate than others. Furthermore, on the whole, the arts themselves are uncritical about the way they present themselves to the public, and without a better understanding of what the arts do in terms of textuality, the understanding of situated knowledge will come to a standstill.

*   *   *

Currently the main tools for discussing the ideology-subject axis come from political and discourse commentators who are not looking at the overall pattern of rhetorical strategy. When we do, we find that the elements that define the stance are embedded in history. The ideology-subject axis comes about partly in response to the larger franchise of the capitalist nation state, and its need for a different kind of government which became representative liberal democracy. Because of the need by capital and nation for stable industry and markets, legitimate power and obligated citizens, that representation was increasingly cast in terms of adequate subject positions. To take representation as adequate helps to underwrite the notion of a perfectible communication: one that can be neutral or pure or offer absolute truth. It is a communication not in need of rhetoric, which is another way of saying that there is only one acknowledged rhetoric, the rhetoric of objectivity. To take representation as adequate also underwrites the constitution of subjects into positions that citizens may take up within the state, so that they become autonomous and isolated from one another. And to take representation as adequate underwrites the stability of nation-state capitalism because there are no alternative communications; change always occurs within the system, never goes outside the grounds of the *a priori*, a reified rational and linear analysis. The nineteenth- and twentieth-century fascination with entropy as a system running down, is an understandable if blinkered fear of people caught in such a rhetorical structure.

Ideology, in representing the state, develops a textuality founded on the concept that language and other media are (in)adequate to reality. Modern natural sciences, which have consolidated into technoscience with the same pressures of nation and capitalism, have also developed a language for communicating to the public which is based on notions of language as inadequate. Indeed, many political analysts take science as the 'best-case' for politics. The following chapters study the way that the rhetoric of computing science and AI often mimics the structure of scientific language without paying any attention to the experiment as the immediate textual practice of the natural sciences. In doing so, and at precisely the moment when they are beginning to provide an transnational language and textuality for global communication, they both underwrite ideological strategy and derail the public perception of the activity of their disciplines.

2 Rhetoric and artificial intelligence

Computing applications in the sciences and humanities

Artificial intelligence (AI), as it has developed over the last twenty to thirty years, is of considerable importance to the study of rhetoric. The field starts with a radical elaboration of formal logic and moves on to a broad expansion of heuristic procedures that introduce context into formal analysis. The movement parallels Aristotle's separation between logic and dialectic, although for Aristotle dialectic should come before logic rather than the reverse. For rhetoric, artificial intelligence can contribute a great deal on technique and device, and a growing amount on strategy. But as yet there is little or nothing on stance, or the centre of rhetoric which is concerned with material interaction. Rhetoric works with a number of techniques and devices, which help to fulfil the persuasive strategies of the writer. And, of increasing interest over the past twenty years, the reader too brings a number of rhetorical strategies to bear upon the text. Classical rhetoric describes these fields under the terms ethos and pathos respectively. At the same time, at the particular moment of history that a text is communicated, the writer and the reader are engaging together through the text, the orator and the audience are engaging collectively in the speech. Attempts to understand that interaction are to do with rhetorical stance.

The lack of commentary on stance from artificial intelligence is odd because in many ways it is fundamentally concerned with rhetoric. Aristotle identified psychology, or the passions, with the field of rhetoric, and it is in psychology that artificial intelligence is making its effective contribution for the moment. More importantly, AI is dealing with a problem underlying the methodology of modern science: how to incorporate context into rational reasoning, which is essential for an increasingly technological science. Artificial intelligence as a field of study is also divided in ways that indicate a rhetorical dilemma more broadly pertinent to twentieth-century epistemology: between S-knows-that and S-knows-how.

COMPUTING APPLICATIONS IN THE SCIENCES

Rhetoric, history and modern natural sciences

Simultaneous with the elaboration of the ideology-subject axis in capitalist nation states, interlinked developments come into play in the fields of rhetoric and of science. During the sixteenth and seventeenth centuries rhetoric in Western Europe completely changed its face. At this time a logic isolated from rhetoric was proposed that was rational and analytical.[1] This may have happened for social reasons because there was an emerging commercial class or group which perceived an education in rhetoric as an instrument of class constraint, which indeed it has often been.[2] It has been suggested[3] that the emerging group underwrote Peter Ramus' attempt to sever logic from rhetoric and then retain logic as the one valid component, generating a mode of pure reasoning. The severance coincides with the development of abstract, typographically presentable geometry, which made possible a self-contained mathematical notation, in turn generating a mode of pure language. This hope for pure reasoning and pure language created an environment in which education in rhetoric had little chance of working fully. The study of rhetoric was demoted from the one essential skill of any citizen taught by Erasmus and other early humanists,[4] to decoration, manipulation and eloquence in oral performance.

But of course, rhetoric itself never goes away: it is always there whether or not it is studied as such. Rhetoric was studied in Scottish universities until the twentieth century, for the reasons of necessary enlightenment and educated social discourse.[5] In England rhetoric was studied less subtly, and with consequently greater loss, than the 'humanities' or classical education. This education did not foreground the pervasiveness of techniques and devices as such; it recognised the persuasion in poetic metaphor but not in scientific topic. The strategies of rhetoric split between the literary and the other: between 'reports', documents and so on which were supposedly factual, and the genres of fiction and verse, the gothic, the sentimental, the lyric, etc. The former take the reader/writer relationship for granted while the work of the latter is to foreground it. 'Poetic' became the term for work that is rhetorically conscious and where some understanding of and approach to stance can be found.

This history is curtailed but may provide some of the historical context for the emergence of modern science in the seventeenth century and its subsequent relationship with rhetoric.[6] The revolution in science that took place during this period saw the public communication of science move from debate to performance. As science became increasingly an aristocratic pursuit, it turned its attention to technology as the traditional field of experiment, not for money but for leisure, and specifically for leisure that could be legitimated as a serious activity. Whereas the sixteenth century abounds in books that tell you how to make after-dinner fireworks to entertain your guests, by the end of the seventeenth century gunpowder, firearms and the like are areas of study about which the carefully chosen members of the Royal Society send each other papers, documented in the *Transactions*. Without the need to be protective of knowledge, in order to gain a living by selling its products, the use of technology which had been referred to as the 'secrets' either of individuals or guilds, could be put into the public domain. For the first time it became possible to compare technologies, to find rules and/or repeatabilities. At the same time the new aristocratic gentlemen scientists were attempting to legitimate their pursuit of what had been the knowledge of specialists who needed to earn money from it, or of domestic workers who produced all that we now so easily buy in shops in order to maintain a household. Simultaneous with the 'democratisation' of science, came the need to show that it was successful, exact to reality, even 'true'.

While there is increasing evidence that many experimental scientists practised at home in their kitchens, when the experiment became repeatable they went to London to perform it for their colleagues. In writing it down, the aim was to enable any other person also to duplicate the experiment exactly. Yet there seems to have been a fundamental belief that what they were doing was not communicable in words, and could only fully be understood by the club culture of other scientists. The structure of the communication of modern science developed techniques to shut out anything that would weaken or question its hold on duplication, and hence its successful evidence for control over nature. This structure lies at the centre of the rhetoric of technology and modern industry, which capitalises on that dependable repeatability. However, science continued to separate between its practices and its communication, the latter being typified by the enclosed tautologous world, a rationalism based on linear sequence and reductive analytics, and a mode of assent claiming verisimilitude or seeing-is-believing.

When the notion of neutral logic and pure language replaced 'rhetoric', it assumed a rhetorical stance which denied that the rhetoric of social persuasion in historical context was needed, even denied that it existed. The techniques of the new persuasiveness necessitated and were built upon the concept of the 'universal autonomous man', able to communicate infinitely replicable experience; isolation from the social world, which might contaminate the purity or challenge the totality of explanation; and carefully structured rational logic. I take the communication of modern science to use the techniques of a rational analytical logic and a denotative language. Its logic moves from premise to conclusion, using conclusions as new premises in a reductive pathway;[7] it can also be numerically described and is therefore quantifiable. Its grammar includes an exact syntax, no poetic, a formal latinate rule system that attempts to parallel mathematical equation. The result is that it communicates in a logic that proceeds toward specific ends and in a language that attempts to be exact in expression of those ends. Its rhetoric claims that there is no need for rhetoric. It takes its grounds for granted, builds self-defining tautologous worlds within which rules can be employed to achieve objectivity. Both the new 'rhetoric' and the new communication of science mimic the autonomous individual of the liberal democratic state, isolated and capable of pure objective judgment, relying on his sight and his capacity for precise, faultless, rational procedure.

Yet within the history of rhetoric, this kind of persuasion is precisely the rhetoric that Aristotle defined as non-rhetorical, and that Plato describes as irresponsible because it is without social placing. From the start of recorded rhetorics, writers agree that rhetoric is persuasion and that it is persuasion from opinions or belief rather than from fact or *a priori*.[8] In other words rhetoric must always first persuade you of its grounds and only then proceed to argument. Plato distinguishes between persuasion from opinions based on unexamined grounds, or the plausible, and persuasion from opinion based on discussion about and assessment of grounds, or the probable.[9] Both are necessarily social, the first showing a careless attitude to political and social conditions,[10] and the second maintaining a responsible position.[11] Aristotle, however, concentrates only on persuasion from probable grounds. He writes the *Topica* specifically about this kind of logic, arguing that it can be used in dialectics as well as demonstrative argument, except that dialectics will also include the topics of practice: of social aspects in the rhetor/audience interaction, while the demonstrative (or aporistic or sophistic) argument will be exclusive of social contingencies. This distinction opens the door to non-social, non-contextual argument.

What these comments amount to is a definition of the relationship between rhetoric and logic. There is persuasion by the plausible, which is an irresponsible evasion of social context, manipulative and coercive: all those things that the popular use of the term now associates with rhetoric. There is also persuasion by the probable, incorporating dialectic, which involves a necessary consideration of social practices and interaction. And we have persuasion by the demonstrative, or sophistical or aporistic, which is exclusive of social context by open declaration. The cautious will note that Plato's irresponsible argument by plausibility is parallel to Aristotle's exclusive argument by the demonstrative. The former is unconscious of evasion, or ignorant; the latter is conscious and its validity, or distinction from ignorant plausibility, depends upon its clear rejection of the social.[12]

Aristotle does not claim absolute truth for factual observation, but he does provide a reasoning which allows us to argue as if there is a set of truths: the key to his proposal is, however, that we may take an opinion as a truthful premise if it is first shown to be acceptable to 'all or to the majority or to the wise'.[13] Futhermore, in *Rhetorica* he suggests that science or physics is a distinct field of study just because it operates on the basis of previously established premises; in other words the only basis upon which we may reason as if there were truths is if opinions or beliefs have already been established by the dialectical argument of rhetoric. Rhetoric is an irre-trievably social argument in which the moral is far more helpful than the true.[14] The social world always underlies scientific procedures and rhetoric, but by convention science operates as if it can take the social and contextual for granted.

This scientific communication was exacerbated during the seventeenth century by precisely that claim to pure reasoning and pure language that was at work enervating the study of rhetoric. The empirical basis of modern science, emphasised and reinforced by someone like Bacon, shifted the ground of scientific activity from the disputational schools into private individual practice. Bacon himself was concerned to incorporate rhetoric into the communication of science so that the public could be told about it.[15] In this he was not being patronising, but recognising that the shorthand, club jargon developed by the small scientific community engaged in this private practice would be elitist/esoteric to a broader public. Rhetoric had the job of attempting to relate these new discoveries to the contemporary society, otherwise there would be no social interaction or responsibility to the public. What Bacon did not do was examine the rhetoric of the private practice of empirical discovery. Yet this became essential when empirical discovery was crossed with mathematical analysis, because mathematics, once transferred to the printed page, became a language that tried sharply to decontextualise the real. Perhaps this, along with the growing influence of a decontextualised logic of proofs, makes it not so surprising that the related twin of plausible rhetoric comes to the fore as the coercive, manipulative definition for rhetorical argument in general.

However, to recall the previous chapter, Plato and Aristotle were also interested in the social and political implications of particular rhetorical strategies. Tautologous and self-directed strategies are the armoury of the tyrant. Aristotle distinguishes between the demagogue and the orator, precisely as between one who persuades without social context and on his own behalf, and one who persuades with the good of the largest number in mind. He further suggests that when demagoguery fails, because it is not engaged with its audience, it must resort to force; whereas the orator does not understand force and can always negotiate. Yet no rhetoric is in itself conducive to demagoguery. The rhetoric of science described by Aristotle is also the rhetoric he attributes to philosophy, and it is helpful to remember that he is positioning the rhetoric within a small group of power holders: Athenian citizens who would have been trained from an early age in many strategies of rhetoric. The Athenian citizen was part of a participatory democracy, fined if he failed to attend and vote on political issues. The difference between this kind of persuasion and the artificial and constrained world of sciences would have been sharp; the understanding that the former was social and engaged and the latter disinterested and abstract would have been clear. Initially it should also have been so to a seventeenth-century aristocrat, but as representative forms of democracy dislocated the individual from political engagement, and as ideology took on the rhetorical structure of science, those people who practised science would increasingly have been educated in only one dominant mode of rhetoric. There would have been little to indicate critique by difference, although there was the continuing study of Greek and Latin as contrasting examples.

But we need to remember that modern science was in the seventeenth century basically a hobby practised by members of the upper classes, many of whom had attended humanities classes in public schools and had imbibed a classical education in rhetoric.[16] Hobbes' development of unitary representation by analogy with mathematical patterns and structures, set up the hopes of the Royal Society pronouncement from Thomas Sprat about scientific language using one word to refer to one thing.[17] At the same time, concurrently and through much of the eighteenth century, people such as Locke, Hume and Smith recognised that science needed a public discourse. But with the advance of technology, or applied science, in the late eighteenth to nineteenth century, this responsibility to public discourse fades. As it fades, so scientists increasingly begin to speak only to each other and to technologists, not to a general public.[18] Their 'models', their tautological worlds, acquired the status of neutral fact. Although it has been argued that modern science has always demanded this status,[19] I have come to think that it is technological application with its particular requirements for commercial exploitation that pushes the wider scientific discourse into claims of neutrality.

It is one of those cultural turns that in retrospect appears inevitable, that Einstein's work on the way that the observer interrelates with the observed and the several extensions of these suggestions,[20] radically challenged the basis of neutral observation and expression of 'fact' just as it was entrenching. But no-one would deny the technological advances that such models/modelling made possible, and the status of modern science remained safe. It is another cultural irony that Russell and Whitehead were at the same time working on their formalisation of symbolic logic, which provided the bridge necessary between mathematics and rational logic that would permit a formal symbolic logic to become the analogy for the neutral, and later 'natural' language expressions of experience and event.

Together, the work on relativity and on symbolic logic forms a background to the development by twentieth-century science of isolated worlds, islands of time and space where the observer does not intervene, and where formal logic can operate in a context-free environment. It is a kind of doublethink: you know that it is only an invented artificial island, but you forget this in order to accept the validity of the logic and language which aim at getting truth by way of the procedure of hypothesis-to-proof. It is important to point out that these isolated islands are not paradigms: Thomas Kuhn's *The Structure of Scientific Revolutions*, which emerged from the late 1950s, suggested that science usually operated within broad ideological contraints to produce 'normal' science,[21] but that at significant moments there were broad shifts, scientific revolutions which built upon a new paradigm within a new ideological set. In contrast, these isolated islands of modern science are rather a plurality of hypothetical worlds, plausible/possible worlds. The validity of their structure is based upon games theory, which was elaborated in the 1930s by Johan Huizinga and then extended in the 1940s to 1950s by, among others, Wittgenstein – but in a rather different way. For artificial intelligence, which emerges as a discipline exactly during this period, games theory becomes an important element.

Artificial intelligence: commentaries on research

The history of rhetoric has always been concerned both with what people have explicitly studied within communication, and with the rhetoric of the way that they have done so. The following discussion will try to address both what AI says it does, and how it says what it does. In broad terms, artificial intelligence can be said to deal with problem-solving events and with knowledge representation, although each overlaps with the other. The discussion of these topics takes place in a public arena, the commercial world, in teaching, in research and in other locations. However, the following study will concentrate on writing for teaching purposes and for purposes of research. Textbooks are particularly important because they are the main way that an ideologically specific discipline turns students into subjects.[22]

It is tempting to rehearse the Hobbes-Sprat-Locke-Hume-Smith chronology,[23] because these were the thinkers who believed they could do without rhetoric in philosophy, although their writing is necessarily rhetorical. Yet despite Hobbes they knew they could not do without rhetoric in public discourse. These thinkers/writers laid out the basis for what we now study as cognitive psychology: what is intelligence and how does it work?[24] This is the question posed by Alan Turing when he proposed the 'Turing' test in the 1940s. The Turing test basically consists of the following proposition: if you ask a machine questions and from its answers cannot tell the difference between those answers and what you would expect from a human being, then the machine could be said to be 'intelligent'. The debate over whether a machine can act intelligently has since become quite complex, and has generated numerous arguments over 'strong' and 'weak' AI.[25] But in a sense these arguments are a red herring, and displace us from more immediate concerns.[26] In any event the Turing test is not really a test for machine intelligence but a test for what human beings are willing to recognise as intelligence. The test works with at least two approaches. The first is that of performance, where one asks the behaviourist question of 'does it act like a human being?', does it say the expected thing? The second and more interesting approach, which asks whether the underlying activity generating that performance has any parallels with human thought, is the cognitive question.

One of the main questions noted by AI people in the 1950s and 1960s was whether it was possible to get machines to solve problems that humans usually solve, and this developed into a field that came to be known as 'expert systems'. From the start AI turned to the premise that thought is 'symbol manipulation'[27] or 'cognitive activity can be described as symbol manipulation'.[28] Because it arose within the computing science environment, symbol manipulation meant mathematical logic, or for those events difficult to code numerically, formal logic and the predicate calculus of if/then logic.[29] Therefore, it immediately acquired an exact language with a rational and analytical logic just like scientific rhetoric. But, because thought was made to correspond to symbol manipulation, which in turn corresponds to mathematical language and formal logic, the very material of AI was exactly made up of exploring the structures, strategies and effect of this rhetoric. In a significant way the early research of AI is a metacriticism of scientific communication, not of its methodology.

Problem-solving AI has by definition an end or a goal built into its structure. It wants to achieve something that a human expert would otherwise have to achieve. Mathematical and formal logic can only achieve goals within the restricted set of grounds they are part of. That we are often unaware of the tautologous structure of these operations, only underlines both the ease with which we accept them and their profoundly ideological basis. In a descriptive language familiar to much AI, N. Nilsson summarised the procedure by presenting the predicate calculus as a language with an explicit syntax, generating legitimate expressions or 'well-formed formulas' – rendered as 'wffs' – by working from ground wffs that are 'true' tautologously.[30] Both 'validity' and 'satisfaction' depend upon our interpretive acceptance of the truth of these grounds.[31]

In rhetorical terms, the points one might want to make are being expressed in a representational language that maps symbol directly on to object and allows one to make an argument that proceeds from accepted grounds to truthful ends. Also in rhetorical terms, to achieve this end one has to insist first on the acceptance of the grounds, most effectively doing so by excluding all others and creating an isolated system; second on the exactitudes of representation, and third on the necessity for or desirability of a goal, an end, a truth. In other words it is necessary to hide the probable or possible definitions at the base of the argument, to deny its rhetorical process.

Such a procedure achieves considerable ends: it is structured to do so. The commercial exploitation of such structures in expert systems is a pragmatic guide to the success of such a rhetoric, as is the wider technological exploitation of scientific desire. The tautological structure of the world allows the computer representation completely to evade the social and political implications of the surrounding world. But it has its drawbacks: it is difficult to change; it is often rigid and made clumsy by the restrictions of its logic; but above all, it is decontextualised. The AI community recognised these drawbacks very quickly.

For example there is Joseph Weizenbaum who, in 1976, describes with uncanny accuracy Huizinga's theory of games,[32] which he never mentions so we must assume is not consciously referring to, and is as beset as Huizinga by the dark implications. Weizenbaum says 'the only certain knowledge science can give us is knowledge of the behaviour of formal systems, that is, systems that are games invented by man'.[33] In order to practise, the scientist must suspend general disbelief, 'must believe his working hypothesis, together with its vast underlying structure of theories and assumptions' (1976: 5) because 'scientific demonstrations, even mathematical proof, are fundamentally acts of persuasion'(15). To do so the scientist must simplify reality through abstraction, and 'abstraction means leaving out of account all those empirical data which do not fit the particular conceptual framework within which science at the moment happens to be working'(127). He also notes computing must adopt a formal language that is precise and unambiguous. Weizenbaum's commentary describes the need for a neutral language and exact reasoning, and in doing so also outlines the limitations: He adds, 'the power of computers is merely an extreme version of a power that is inherent in all self-validitating systems of thought' (130).

The positive aspects of such an epistemological strategy are found in the way that such thinking sharply indicates the limitations of the tautological world. A model is useful because at some point it fails, and in that failure it makes a comment upon the real. The strategy is also positive for the way that it provides metaphors for thinking about things in new ways. But on the other face of these activities are the negative aspects which arise from setting up truth as an achievable goal. Such success gives the games player, the programmer, a sense of enacting power, in other words a sense that the tautological world is the actual world, and that there is no failure, no limitation. As a result the game rules become laws that effectively shut down new ways of thinking, and shut out social and political immediacies.[34]

In the 1930s Huizinga sets up games as social systems, clubs, necessary coteries with end-directed rhetorical strategies, and he happily makes science parallel with this activity. Yet after World War II he recognises that his club culture exactly described the rhetoric of Nazi propaganda, and his conclusion to the 1949 edition of *Homo Ludens* voices anxieties similar to Weizenbaum's on the tendency for such worlds to become total. Weizenbaum adds the further terror that computers may be being made with the aim of replacing human beings.[35] This of course is the popular misconception, yet it is surprising how often one finds this aim implicitly asserted in serious AI literature. From a rhetorical perspective, the problem with these games is that they become conventional; they produce a hidden rhetoric that is dangerous because if we cannot assess it, we do not know how to deal with it.

A less prophetic and certainly less political analysis is made by John Haugeland. In a book specifically related to AI, he documents the history of materialist thinking on cognitive psychology, to the end of the eighteenth century.[36] The commentary picks the history up again in the twentieth century with the emergent field of artificial intelligence, conveniently leaving out the nineteenth century as 'idealist', yet in the process also leaving out Marx, who supplies the one thing Haugeland's argument is trying to move toward: context. Haugeland reiterates the necessity for computationally based reasoning to abstract its arguments, to the point where electronic and physiological structures, machines and bodies, are on the same footing (1985: 5). Because reality is mathematical in such abstraction, it needs a formal system to describe it. The formal system is isolated like a game, it has a digital language where the token equals the thing/referent, and it has a rational logic of deterministic algorithms. After much intense analysis the commentary concludes that this reasoning only works in axiomatic or tautological systems like mathematics (107–9). Furthermore, it only appears to work in computing, first because of the scientific conventions that underlie computing strategies, and second because there is no context and therefore no audience. It is exactly at this point, Haugeland argues, that AI can step in and try to deal the computation of interpretable or semantically significant events. In an effort to address context, he suggests heuristics, he discusses General Problem Solvers, expert systems, stereotypes, frames, as methodologies that have been put forward for contextualising formal logic. However, since he has been at such pains to prove that heuristic procedures and other methodologies can be reduced to deterministic algorithms (83, 117), there are drawbacks to his discussion.

The description in Haugeland's commentary illustrates a claim from Hubert Dreyfus that when the AI community had pushed the limits of formal logic as far as they could go in the mid-1970s, they turned instead to looking at ways in which context, what they blithely refer to as 'commonsense' could be introduced.[37] In many ways this was the turning point away from Huizinga's theory of the ideological basis of games for large social structures, to an interpretation of Wittgenstein's comments on games as individual strategies or designs.[38] In the process there has been a general 'turn' to questions of 'natural' language. Weizenbaum suggests that the shift was based on a belief that computer programs failed because the language of formal logic is limited, hence if we use natural language there would be more potential for success. But the question became: how to describe natural language in computational terms?

Noam Chomsky had spent the better part of the 1960s arguing for innate, biologically determined grammatical structures which could provide an alternative to formal logic.[39] It was important for AI that Chomsky claimed biological determinism for his grammar, because it provided a firm foundation for the basic assumption that thought itself, or the brain processes in cognition, is symbolically structured as an essential. To quote from Yorick Wilks, Wittgenstein convinced 'a generation, already intoxicated with the power of formal logic, that logic was not necessarily the structure underlying thought and language', and Chomsky went on to provide 'a systematic structure of forms that were not those of formal logic...some real systematic alternative candidate as the real structure of language'.[40]

Chomsky had also taken the separation between deep structure and surface structure, that was made but not defined by Wittgenstein, to develop transformational grammar. Despite the fact that Chomsky's grammar was itself non-contextual, AI found that it could use transformational grammar as a means of moving from the innate to the contextual by developing semantic networks.[41] Indeed, many of the discussions of structure in AI research between 1975 and 1985 turned to concepts of generative and transformational grammars. Another influential direction also emerging during the 1970s was Minksy's development of 'frames'.[42] The strategy resembles a simplified version, simplified presumably for computational reasons, of Gerard Genette's theories of narratology; and it too aims primarily at introducing context via the concept of stereotyped knowledge.

What such movements indicate, and I would not want to suggest that formal logical was being dismissed for it is still the primary strategy of AI,[43] is an acknowledgement of the complexity of representation through language, and therefore of the complexity of the representation of knowledge that results from the need to include context. We may see a similar discussion going on in the development of nonmonotonic logics in their search for temporal and spatial contexts,[44] and in the explicit concern with metaphor as a way of representing knowledge.[45] In AI research we can watch a swing toward far more subtle considerations of the question of knowledge representation, which of course is necessary to its counterpart, problem-solving. Yet in this quotation from 1986 the aim is still toward 'explicit structured representation of the underlying rules of human expertise'.[46]

Again, the AI community quickly recognised the limitations of natural language representations, and has performed the same metacritical exercise for the scientific aims of a form of linguistics, as it did for modern science as a whole – although neither metacritical commentaries have been fore-grounded. For example, Drew McDermott has spoken eloquently of the limitations of using tools based on natural language patterns, but perceives the fact that these are a main tool for linguists as an advantage.[47] This is despite the irony that few linguists exactly use Chomsky's grammars, precisely because they are so decontextualised.

Recognition of context has meant a recognition of games-design as opposed to game-rules for specific applications. It has become a way of defining a very small 'possible' world into which some elements of context may be introduced without destroying the system with over-complexity. In other words, games-design is about creating a discourse within which context can be partially formulated. It is not about creating a discourse that deals with the impossibility of formalising context. The distinction is significant, for while it has been a major step forward to move from technique or device to strategy, what games-design does not take on board is the ease with which such strategies shift into the fixities of formal ordering over time. One of the clearest outlines of this shift may be found in J. Fodor's cheerful commentary on AI as 'methodological solipsism' or tautological worlds.[48] He notes that formal logic does create the tautologous worlds typical of solipsism; just so do the designs that incorporate predictable context slip through habit into formality. As Haugeland suggests, all heuristics can be reduced to algorithms. Fodor goes as explicitly to outline the necessary limits of mental representation as a methodological solipsism that has to exclude reference, semantics and truth. Dreyfus argues that even if you could translate a formal description of the world via rules, even if you could specify a context, it would only be one context and would immediately change.[49] He concludes that AI is committed to ignoring the non-cognitive; specifically it is committed to ignoring our bodies (204).

This warning comment from Dreyfus, with others such as Weizenbaum's suggestion that there is a need to understand human reasoning as a function of physiological bodies,[50] indicates a direction that AI has recently taken into connectionism and neurobiology/physiology. As if once more on the search for the formalisation of thought in a symbolic representation, AI has turned to the human body and particularly the physical brain, although other parts of the body are studied for robotics, hoping to find grounds for cognitive psychology in the place that appears to resist it the most. In doing so, AI has generated connectionist representation which is supposedly even less formal than grammar but is, importantly, still modelled on a biological determination and hence still holding out the promise of a specific answer to questions about thought.

A collection of essays from Daedalus that came out early on in the field of connectivism, indicates the utopic desires that have been placed upon design. For example S. Turkle enthusiastically outlines the way that 'computers provide sciences of mind with a kind of theoretical legitimation that I call sustaining myths'.[51] In a Fodorian tone, Turkle interprets the 'mathematical culture [of computer science] that relies heavily on defining things in terms of themselves' (264), as a positive myth, legitimating self-analysis and referential epistemology of 'objects'. The interpretation underwrites the desire for game-rules in much AI, and completely inverts its metacritical possibilities into the fixity of 'sustaining myths'. However, other contributors to this (Graubard 1988) collection indicate a general recognition of the limitations of design. J. Cowan and D. Sharp outline the traditional AI strategy as one that is not context-sensitive. In contrast the designer of the neural nets of connectivism only needs 'informal understanding of the complexities of the desired behavior'.[52] Yet despite this, 'All progress to date in the soft-wiring of nets to perform intelligent tasks rests on prior analysis by a designer of the context-dependent tasks to be performed. The intention and the meaning are supplied by the designer' (114).

The most helpful, if typically contradictory discussion, is that by H. and S. Dreyfus, in which they identify the ability of AI to provide the forum for the invention of small possible worlds with Husserl's attempts at 'top-level sameness' resulting from a shared belief system.[53] They go on to examine Heidegger's challenge to this context-free rationalism and his proposals for 'circumspection' and social construction. While this latter may provide the motivating analogy for the development of neural nets so that they 'may show that...we behave intelligently in the world without having a theory of that world', in practice, to avoid ambiguity, neural network modellers 'design the architecture of their nets so that they transform inputs into outputs only in ways that are in the hypothesis space. Generalisation will then be possible only on the designer's terms' and the net cannot 'adapt to other contexts' (38). But to conclude, Dreyfus and Dreyfus once again turn to the body, and warn that the nets, to achieve the utopic wish-fulfilments of their modellers, 'must share our needs, desires, and emotions and have a humanlike body with appropriate physical movements, abilities, and vulnerability to injury' (39). It is generally agreed that should neural nets be able to be modelled to such sophistication, it will not happen for a very long time. In AI terms the nets have already failed, although as a metacritical study of neurobiology they still have an immense amount to contribute.

Artificial intelligence: textbooks

Virtually all the research writing in AI is aware of the limitations of formalising, computational abstraction. As a genre it can speak with profound understanding and eloquence of the problems and restrictions. Nevertheless, the community is still striving after a neutral or pure representative language, after a logic that will cope with testing hypotheses, and after a restricted definition of the world. The research still believes in the possibility of an expression or representation that will be adequate, will achieve a truth. Therefore it is still denying rhetorical necessity, which is itself based on denying the possibility of end-truths and concentrating on contingency, and on working at valid action in the face of this. Despite the immense contribution to rhetorical studies, particularly on the activity of so-called 'neutral' or pure languages and logic, AI discusses them as failures rather than as necessities. This underlines the residual desire for just those neutralities and places the research in the position of plausible, coercive rhetoric rather than the openly demonstrative. But we need to look at the possibility that this is due to the kind of rhetoric that is often used within relatively closed communities, where much goes unstated because it is tacitly understood.

The presentation in the educational field that complements research writing is the textbook. The preceding discussion is a fairly prevalent intellectual debate within the AI research field. It indicates some of what members of the research community say to each other about the promises and limitations of their work. But what, it might be worth asking, do they say to their students? Do they want to hide or discuss the implications of an approach to problem-solving in terms of tautological worlds, and to knowledge representation as exact and accurate? Textbooks, as P. Winston says at the start of his widely used textbook *Artificial Intelligence*, are very different from the research field. But exactly how do they differ? With students, who are by definition coming into the community from the outside, one cannot take tacit understanding for granted. Winston's own book, a product of the 1970s, is formally based and makes considerable claims for the extent of problem-solving.[54] N. Nilsson (1982) is more concerned with a generalisation of the broad range of AI methods, but at the same time shows more overtly the aim at 'control' that AI desires. His tripartite summary of production systems, data operations and control, refocused as declarative, procedural and control knowledge, is familiar to many AI textbook introductions.[55] Just as familiar is his placing of the formal symbolic logic of the predicate calculus at the centre of his scheme.

During the 1980s, one finds Michie and Johnston's *The Creative Computer* (1984),[56] and Charniak and McDermott's *Introduction to Artificial Intelligence* (1985)[57] making similar claims, but with a growing awareness of complexity that moves them to approach the issues with significantly different rhetorical strategies. Michie and Johnston present AI as providing knowledge or 'the capacity to give correct answers to questions'. In other words knowledge is information, not wisdom,[58] in the classical construction of modern scientific rhetoric. The book claims that the central challenge is the construction of a mechanical logic of common-sense meaning (17), because computers must be made more like humans, not only so that they can be treated as servants rather than slaves, but also so that they can better be controlled.

To effect this movement to common-sense logic, the book starts out with a radically different and potentially valuable approach that provides a series of examples from art which are redrawn in order to give computation access to creativity: artistic, religious, political and emotional. Yet their 'creativity' is merely associational (32); thinking is simplifying or 'lying' (92). The redrawing goes in for hierarchies of creativity which set the small, domestic creativity against the artist as hero (129). This version of 'art' makes no comment whatsoever on aesthetic theory, on representation as necessarily deceitful, on writing in particular as interactive (156), or of the alphabet as a medium;[59] let alone any reference to reader-response theory, psychoanalytic concepts of the subject, the ideological construction of cultural artefacts, or the debate surrounding structuralism. As presented here, the concept of 'art' is grossly naive. For example, in a discussion of classic versus romantic aesthetics, the authors are writing under the delusion that romanticism engages in innocent perception (161).[60] The book's introduction concludes by worrying that philosophical questions may confuse our commitment to 'exploit' the potential in AI for power and control (113). Indeed the entire text is underwritten by metaphors of exploitation, pornography, colonialisation and heroism.[61] There is a sense that in the context of Michie's other work, parts of this book could be read as a parody. However, parody by definition requires considerable prior expertise in convention, which a student approaching this discussion of AI would be unlikely to have. This is particularly so since many AI writers use the narrative structures and vocabulary of the literature of science fiction and fantasy, often drawing on these books for analogies;[62] but they seem unaware both of the double-edged rhetorical activity of the genre,[63] and of the current domination of its exploitative strategies.

In contrast, Charniak and McDermott, in writing an all-encompassing textbook that has more in common with earlier teaching texts, are unexpectedly but also superficially more conservative. Even these writers, highly intellectual and subtle philosophers, choose to introduce AI to the student as, 'mental faculties through computational methods',[64] saying that the 'brain at some level is computation'. The book goes on to present a central contradiction, saying both that axioms will change over time (20), and that 'understanding' is the ability to deduce the correct fact (20). What is lacking is any suggestion about the relationship between deduction and axiomatic shift, or any discussion about the social and historical pressures beyond. Later on in the book they attribute inanimate causality to emotion – for example 'guilt' is supposedly natural – and they make no mention of ideology (568). This AI textbook makes the familair claim that the predicate calculus is the most flexible vehicle for capturing context (320). The writers get rid of the objection to the predicate calculus on the grounds that it cannot deal with people being inconsistent, by saying that the predicate calculus is all right because people in any case always try to 'explain away' inconsistency (344). They get rid of the objection to expert systems on the grounds that they are too powerful, by reiterating the fact that these systems have 'no great insight' (437). Curious comfort.

When Charniak and McDermott's textbook moves on to discussing language systems, it claims what many AI textbooks claim, that writing is simpler than the oral and therefore more open to computation (175).[65] This is a misunderstanding of the medium that is all the more problematic because the written, or a restrictive set of it, is the medium of the grammars that have been taken up into computation. However, the approach allows for a confident illustration of how the rules of formal logic work well with the designs of declarative sentences in grammar (582–4). They are mapped on to the designs of performative sentences in speech act theory, which simultaneously excludes any discussion of the implications of this double act, or recognition of the intimate compromise designs make with rule-bound systems. The conflated perception of written or oral verbal texts also allows for a simplistic approach to the functions of metaphor and analogy,[66] that underlies one of their broadest claims: that AI will be able to provide more, and more frequent, Kuhnian paradigms (650). The mapping of the technique of analogy on to the concept of a paradigm as a model, shows no overt comprehension of the understanding of the radical separation between an epistemological set with a positioned context, and the closed tautologies of possible worlds. Charniak and McDermott are deeply involved in the doublethink: the need to create possible worlds and to simultaneously remember and forget their artificiality. The point here is that if writers such as Charniak and McDermott are not foregrounding the rhetorical necessities to their students, then the discussion about promises and limitations has little chance of reaching the public or commercial world.

Story

Auto(bio)graphy is necessarily selective. We re-member the past into bodies recognisable to the present that carry our future forward. When I re-member my past work in science, a painted medieval landscape appears, slightly two-dimensional with sequential panels of stasis but little or no temporal movement. It is possible to see myself at sixteen, eagerly proud to have won the opportunity of working in a lab in the local university – how difficult to veil the eagerness, the thirst for adult knowledge, legitimate and respected knowledge. At that job, the young woman washed erlenmeyer flasks all day and every day, sometimes twenty to thirty times to ensure contamination-free glass for growing tobacco-pith cultures. The incredible tedium did not appear to be malicious, but it was as if there were a permeable glass wall between what she did and what everyone else in the lab was doing. Every hour or so she handed trays of flasks through the glass membrane to the other side where people needed them for parts of the routine, the choreography of repetition that patterned the movements on the other side into purposeful action.

Because no-one spoke to her, she began to read stories about the other side. There were textbooks in biochemistry and, in the coffee room, journals and magazines about science, that peopled the landscape and offered a vocabulary for talking about what those other people were doing. So that one day, when someone finally spoke to her and in scientific language, she found she could answer. What is memorable about this stationary moment is the responding gesture of the questioner: neither approving, nor dismissive, nor sarcastic, nor congratulatory, it was a banal gesture of cynical recognition that only now do I see as a welcome to the club. Nothing changed. She still washed flasks every day. But the glass membrane moved over her, extending its filaments like optic fibres into the musculature of her body, passing to the opposite side of the sink so that the cleaners, who had always been part of her world, became part of another.

Inside the glass womb of the lab there were still strangenesses and levels and rules for the routine. In fact it was far more difficult inside than outside, because inside the glass the static panels were mobilised, all the people became filamented into figures in an epic drama from which there was no escape. The only respite was the coffee room, in which by an unvoiced agreement everybody behaved as if they were in a familial community. The young woman concentrated on taking the sugar out of her coffee, a grain less every day, till she was down to two grains (granulated) and had become a topic for discussion. The comforting familiarity of the coffee room was a domesticity without responsibility, a refuge from the quest and a sanctuary from aggression. They are all the same – no matter how different the plastic trim, coffee rooms become thick with the film of everyday trauma, the detritus of magazines in whose display our bitty lives are felted together with impressions of coherence. She would construct fantastic designs from the pictures and go home to sew them into helplessly inappropriate statements of elegance and worldly strength that made her real family laugh.

Outside the coffee room was a landscape of hope, desire and ambition. Like types in an early allegory, people would put on their armour and stride out across the plain of microbiology: the director of the lab a benevolent king beyond approach or reproach, yet also, secretly, the lurking dragon in our lives, the one who already had the treasure. His knights the research scientists, some questing for their own treasures, some there for a fight with incalcitrant facts, some half-stunned with reality, all running up the bridge of hypothesis and down the slide of results, everything an object of desire, a grail promised by the game. The young woman spent three years inside the glass womb, learning the rules slowly but learning them merely to survive, not understanding that only those with the privilege of risking their lives can control the game. Seduced by certainty, she consumed textbook after textbook, trying to comprehend the strengths of her newly filiamented body.

And she too was an object promised by the game, a certainty to be seduced by others. For most of the (mostly) men working there, the ethos of romance quest so permeated the world that anything, including each other and obviously including those lower down the levels like her, became fair game. Season after season the research scientist in her lab harassed her, hunted her, laid seige; and season after season research assistants rode in: one chivalric, hoping to defend her as a lover; one working to a different cultural code, hoping to defend her as a wife; one engaged in continual class warfare, as a friend; and one, who did not quite belong inside the glass, bringing, I remember vividly, his wife Lydia to help the young woman to remember what was outside – vivid Lydia offering a different life.

The glass womb had its own rules for science and class, science and age, science and gender, and science and sexuality. I would recognise only later the rules for science and race, science and ability. It was not unusual for a young woman to fail to understand the strategies of harassment and containment as invasive filiamentation. They were similar to the rules that to her, at that age and time, bound science itself. They were the rules through which she understood her own body. She acquired a vast pool of rote knowledge of science, the more successful the closer it got to mathematics, which has its own comprehensive language. Physics always eluded her, and I think now that physicists gave up centuries ago on language in their quest for direct contact with the real. She learned by the book and was tested by the book, while practising the craft in an entirely different and tacit manner.

Another stationary moment of re-membering occurred at the start of university, when the young woman learned for the first time that science was not exact, that it could not be learned by rote. Cursed with a photographic memory, she failed a biology exam on the basis that she had reproduced too neatly and exactly the words of her tutor and lecturers; she had not applied them in response to the material demands of the questions; in effect she had failed to understand that science is textual. An undergraduate degree should teach this, as it moves you from repeating the basic experiments, the building-block interactions with the natural world that lead us to acquire a sense of social behaviour toward it, to specific interactions with it. The work teaches the language of scientific communication, through a variety of technical interventions, observational methodologies and mathematical narratives that introduce you to the etiquette of western man's relationships with the physical environment. They begin to teach you how to control the rules, how to work with them and how to change them. Depending on the etiquette, which is after all all we have to guide us, there is more or less certainty, but it can never be learned by rote.

The surprise came when she left the worlds of science for the arts and humanities, only to find them in the middle of a scientistic revolution. They valued her learned 'scientific' ability to conduct structuralist analysis, or morphological genre differentiation, or numerological study of syllabic patterns. But out there was no glass womb. The figures in the landscape thought her results were 'true'. They did not recognise the rules of the game – indeed they called it 'reality', and in reality there was no coffee room, no refuge from aggression unless one retreated to the isolated spaces of pages and hid among the printed words. Little by little she came to understand that the invisible walls of the arts were part of the invisible walls of ruling political power. Most of the people around her had never been outside that membrane; some did not even know there was an outside. Yet they stared continually at the interface, wondering at its sudden eruptions and its apparently spontaneous generation of strange and different forms. They sank their hands into its fleshy density, trying to comprehend how it worked – but were far less aware than the scientist of the filiamented infusion of their epic lives.

COMPUTING STRATEGIES IN THE ARTS AND HUMANITIES

What is interesting is the way that so many textbooks on artificial intelligence gesture toward the arts and humanities in a recognition of the need to provide context, while there has been so little detailed study of what precisely the arts might offer. In AI and related computer science applications the knowledge being represented, the reality being modelled, is scientific. Despite attempts by AI to introduce context into rational analysis, most of the textbooks which are the training grounds for the discipline's subject formation, focus on linear, rationalist, predicate calculus. Yet surprisingly in the face of the gestures toward arts' strategies made by many in AI, the humanities' use of computing applications from the 1970s to the early 1990s also focused on scientific methods, especially hypothesis formulation and testing by mathematics and statistics. The following discussion offers a series of examples from the field of literary and linguistic computing.

Over the last hundred years of English studies in universities, literary criticism has frequently tried to legitimate itself by looking for methods that paralleled those successful in science. It has been those areas of criticism that had previously attempted scientific legitimation for their knowledge which first moved into computing. Fundamentally, these areas were concerned with database organisation: bibliography and publishing history, bibliographic analysis being the factual basis for the history of printing and publishing; and textual editing and literary stylistics, the former being an extreme mode of the latter. Significantly, both areas have a history of attempting the scientistic, with aspirations to neutral observation, objective knowledge, and precise and accurate truth. Analytical bibliography aims to describe a copy of a book so accurately that it can be defined as a copy from a particular state of an issue of an edition.[67] And textual editing began to receive serious attention only at the end of the nineteenth century, when it began to strive to represent a fixed text as the definitive edition. Both areas of study work within closely defined scholarly guidelines that until recently had extremely narrow concepts of admissable context. Although it has to be noted that during the past twenty years this narrowness has been challenged, and many people working in these areas are now attempting to use them specifically to open up the social and historical contingencies of literary texts.

At the same time, all people working in the humanities organise data. It is a strategy at the centre of their activities, but is modulated by different rhetorical approaches. The differences were sharply recognised by the social sciences during the 1960s and 1970s, and became a debate between quantitative and qualitative methodology. Everything becomes data[68] once we start questioning it because our questions impose a structure or organisation whether or not we remain fully aware of it, and as a species we seem dedicated to searching for pattern.[69] The classic, western, post-Renaissance formulation of rational activity is to conceptualise fact, or the material at hand, being turned into information as we describe it, into data as we order and organise it, and from data into knowledge as we interpret and understand relationships within the ordering structure, or database.[70] Data for the computing scientist tends to be a relatively unproblematic collection of information, and a database is an organisation of information so that it is non-redundant. But at root the problem for a humanities person is that the humanities are concerned with the process of turning information into data, and data into knowledge, as well as with any knowledge that may result.

Furthermore, for the humanities the concept of redundancy in verbal texts is extremely complex. The differentiation between fact and information becomes complicated once we question how perception of the material can be separated from description, and that between information and data collapses once we recognise the ordering structures implicit in our descriptions, in other words, once we become aware of its textuality and rhetorical structure. Understanding, or our conscious awareness of the structures propping up or standing under interpretation, becomes synonymous with recognition of those structures, and knowledge becomes tautologous, self-defining, systematic.[71] In literature and language studies, most practitioners know that if they take a pre-made pattern such as Northrop Frye's theory of genre, they can make it fit; but there is far more interest in the patterns that do not fit, that continually release new significance. Workers in the humanities are not used to a database which does answer questions, but to a text which insists on context and hence instigates rather than answers questions.

Another issue comes from the fact that computer databases are so often interpreted by numerical analysis, and most students in the humanities do not know enough about numbers, their limitations and their gifts, to be able to use them constructively. Some humanities users wield number as though it were a nuclear warhead on the end of a rolling pin; while others see that rampant crudity and object to it. These responses occur for good reason. Mathematical data often generate a neat, tidy and finite body of answers: the knowledge it allows us to accumulate can be perceived as exact. But this exactitude depends upon a circular argument. You start with your axioms, apply them to specific observation/information about a problem and generate data which can be processed into answers. The answers often look incontrovertible or factual, which is the word that has come into use for this kind of knowledge.[72] Those 'facts' or answers then become the premises for more information to be processed into more data to be interpreted into more (factual) knowledge.

The activity is precisely that of the systematic procedure for rational and analytical ordering outlined above. It is this procedure that has been at the base of the criticism of modern science as reductive. The reductiveness produces an inverse tree structure where the branches are ignored and the analyst moves further and further from the actual world. The tree, rather than getting bushier and bushier with the contradictions of the real, gets thinner and thinner and further away, until we forget that the axioms are only axioms rather than actual things in the real world. Numerical analysis easily shifts into an insistent rationality which can carry touches of monomania in its curious inexorability of movement.

The cliché introduction to databases and computers is that they are number-crunchers and somehow more suited to scientific studies. Indeed, advances in computer development have occured when large amounts or a large number of details need to be controlled and analysed, for example in the 1890s when the United States decided that it needed machine computation to help it with the sheer mass of information in the census.[73] Looking at the census we can get a good idea of the extensions and restrictions involved in an early computer database. Here the applications are not just dealing with numbers; with the census we are already into the untidy area of people. What does the census do? Its printed form is parallel with its computerised form: specific kinds of information that are gathered about people are presented in columns (printed) or fields (computerised). The presentation implies information that is analogous from entry to entry so that addresses, occupation and so on can be compared. But how were the categories chosen? how was the information obtained? how was it interpreted? What does one make of a person entered as a 'housewife'? Is this her description of herself, or someone else's? What does it signify? does she have children? Does she work part-time? Does she live in a four-bedroomed or one-room house? In other words where is the context, and what has been omitted? When you are first introduced to someone, you rarely gain much understanding of them even if you ask the most direct personal questions – a lot is left out. Much more is left out of a computer database, and what is put in guides our interpretations in a more strategic way.[74]

Degrees of database organisation

Simple literary texts

In a sense the textual critic, or educated reader, inevitably turns a book into a database by way of the reading and writing skills they have learned. Frequently the reader is looking for a pattern in the text, even if it is a pattern that disrupts or is disrupted. To recognise a pattern the reader or critic will look for discernible elements which are taught as part of an education in the literary: an understanding of lexis, of grammar, of communication or rhetoric. Theories in literature and language criticism often arise as attempts to suggest ways of looking for different elements: for example, the now rather narrowing pairing by Jakobson of metaphor and metonymy, or Genette's narratology, or Booth's irony, or Halliday and

Hasan's cohesion.[75] Questions are usually asked because the critic is interested in the topic, in context, in social and historical relations. Studies of vocabulary, figures, structure, all underlie studies of context and topic – although they can have no significance divorced from them.

Given that most critical readers can point to elements that they take to be significant and explain why they are focusing on these and asking questions about them, the computer has been able to help in a number of elementary ways with far-reaching implications. First and foremost, computers are good at finding specific sets or strings of letters, of words and groups of words: single or multiple instances of them, or occurences of two or more together or close to each other. They can search quickly for significant clusters of letters words or phrases, and their co-occurences or collocates. This is quick and easy to say, but the implications of such concordancing are vast, for they include searching for rhyme, alliteraton, for connectives, for the location of accumulation of similes (like, as), for common words, for proper names, for amount and distribution of potentially significant vocabulary, for type/token ratios, for the use of exclamation or apostrophe, for dialect or slang words. Many of the early databases constructed for literary and linguistic study selected significant elements right out of the text and placed them in predetermined fields to make it possible for the computer to perform matching and comparing operations.

Most of the early work in this field was done in authorship studies.[76] Its often-naive approaches may be recuperated as constructive, both through the social context they can make relevant when for example one is suddenly able historically to locate a piece of writing, and through the internal evidence of the use of vocabulary and syntax specific to a particular voice in a general historical setting.[77] For example, a reader may want to have a quick look at the use of the words 'good' 'honour' and 'value' in Fielding's *Joseph Andrews* in the light of the writer's essays on morality and behaviour. With careful and intelligent thought, and considerable criticism and rejection of initial proposals, the computer can relatively easily provide you with a list of the occurrences of these words and cognate forms, with some context around them,[78] from which it is possible at the least to deduce whether the inquiry could have a large or small amount of material to work on, and more positively it might also be possible to begin to see patterns of textual use and strategy.

Structured literary texts

When looking for groups of letters, words or phrases, the text is being organised into data by the structures of lexis grammar, logic, rhetoric and poetic, which the critic has learned and which can be brought to and found within the text without altering the presentation of its traditional printed form. The guides to structure are primarily the order of the letters in words or sets of words, from which order the computer can locate, count or compare. However, once the critic wants to move into more sophisticated questioning, searching for significance which is not carried by the order of letters, there are different requirements. On the one hand it is necessary to invent a way for computers to recognise not only letters but also narrative elements and rhetorical devices, and on the other hand for some way of indicating such aspects with special markers in the text. The latter route is relatively simple but time-consuming. The former route is complex. Currently it involves many students and scholars in natural language processing and artificial intelligence,[79] and has come under increasing criticism for its dependence on a reductive, decontextualised symbolic logic.[80] Yet even the former route, which lies closer to current critical practice, is not without its potential for losing context.

A clear example of a set of texts marked up for specific study would be John Burrows' computerised versions of Jane Austen's novels.[81] These are marked up to indicate to the computer not only larger aspects such as chapter divisions and characters, but sections of the text that Burrows has designated as narrator's voice, character narrative, and character's speech. The separation between character narrative and narrator narrative allows Burrows to carry out some extensive study of how Jane Austen constructs a 'voice' for the characters' internal speech or consciousness as well as for their spoken dialogue. Burrows also codes many elements from the level of proper place names, to the level of homonyms such as 'this' as a determine and 'this' as a relative pronoun.[82] To do this he has not only special markers such as asterisks and dollar signs in the text of the novel itself, but he also allocates each line of his computer text a reference field indicating the larger elements such as chapter divisions.

These organising structures are the result of the kind of activity which is recognisable to any working critic of literature and/or language. All Burrows has done up to this point is simply make the detail of a long text open to the kind of analysis we might usually expect to give a short text, by coding the words in such a way that the computer can quickly scan them. But these structures are also those which foreground some of the ways in which we often unwittingly impose database organisation upon what we read. Burrows would not claim that his structures answer all questions; they are not intended to do so. They are there specifically to assist his pursuit of certain questions he wants to explore.[83] That other students of Jane Austen's work may benefit from his creation of an analytically marked-up text is in a sense fortuitous, and results from a common critical consensus about significance. What his activity does do, however, is make clear and visible some of the operations which many of us take for granted. He leaves his methodology open to scrutiny of a particularly acute kind, since it is all too easy to question not only whether his choice of where significant linguistic or narrative events occur, but also whether it is helpful to think of those events as significant in the first place.

Burrows' activity in creating a version of Austen's novels as a database is directly analogous to that of a textual editor creating an edition for publication, and indeed to a sociologist of language preparing a transcript. The textual editor has, since the start of printing, devised a set of rules of thumb different from those of manuscript copyists, and appropriate to the reception of certain printed texts in specific historical periods: Pope's version of Shakespeare is quite different from say the nineteenth-century Bowdler edition or the contemporary Arden Shakespeare.[84] Just so the sociologist of language has had to radically question the presentation of reported speech and communication, and devise a methodology appropriate to its topics.[85] Since Burrows' work in the 1980s, several large publishing houses have followed up the possibility of using the computer to provide added social and historical context to editions. A rather different and more radical approach to textual editing by way of the computer is taken by H. Gabler in his edition of James Joyce's *Ulysses* (1985, Penguin). Here he departs from the usual aim of textual editing to select and make accessible significant elements of the text, and instead uses the computer medium to present a complete text of all six editions of *Ulysses* published during Joyce's lifetime. No context is offered outside of the dates of the specific segments of text, and the responsibility for selection is placed in the hands of the reader. Yet at present, without wider understanding of the ethical and historical implications of textual editing, that responsibility is difficult to fulfil.

Selective texts

Even more overtly organised databases are produced by bibliographers, because they are building models of knowledge different from the sanctioned and inviolate 'verbal text' of the artist-author. Partly because of the difficulty of studying elements abstracted from textual context without a nervous worry about reductive narrowing, and partly because for centuries work in bibliography had been trying to restore a non-verbal context to literature and language by means of overtly organising techniques analogous to those employed by computer strategies, many early computer-assisted studies relevant to verbal texts were carried out by bibliographers interested in the social and historical context addressed by the printed medium for communicating words. The classic database in bibliography is the 'flat file' database,[86] which has the same type of information on each card, and the same fields for each entry. They assume that the distinct fields of information will not change, and they can be adapted for a number of different database questions, or programmes, relatively easily, because a lot of work is done by the bibliographer in selecting and defining the significant material. They also, for this reason, work quite fast in computer terms: essentially the programme, or set of possible questions that one chooses, speedily matches up previously defined fields. The main thing that makes one programme different from another is the way in which it can match up different fields.

An example of this kind of database, and there are many, would be one that a group of researchers (Attar *et al.* 1987–9) created of domestic texts published in Britain between 1800 and 1914.[87] There are around 3,000 separate individual texts, and roughly 25,000 distinctly different editions of the text, so 25,000 different index cards would be needed. This represents a substantial amount of description. Further, each card contains fields for author, title, publisher, collation, description, citation, location, notes, topic, and more than forty subfields. The database was initially set up for the FAMULUS[88] database structure, which allows for swift indexing and which was particularly helpful for generating chronological, short-title and topic indexes. Because so much work had gone into marking up different kinds of potentially significant detail, the database was also adaptable to other structures such as STATUS or EXTRACT,[89] which allow questions and combinations of questions about, for example, which authors were publishing with a specific publisher between 1860 and 1870, or the average cost of domestic books published in London during the 1890s compared to those published in the regions.

This type of selective bibliographic database is a powerful research tool, and many are now available.[90] What they offer impinges directly on questions of genre, audience, authorship, copyright, censorship and printing history. Provided that care is taken with the decision about how to organise information, this type of 'flat-file' database can be relatively flexible and helpful not only to bibliographers and literary critics, but also to historians of publishing, society, economics and culture. Its limitations lie in the historically specific necessity of selection for significance. In a sense it leaves itself even more open to criticism of its selection procedures than a computerised textual edition because it omits so much; at the same time, simply because of years of bibliographic tradition, there is more agreement on the kinds of description appropriate to the audience the bibliography is trying to reach. A flat-file database can of course be constructed for a literary text as well, but as noted it leaves out so much that critics used to dealing with textual detail often object. Nevertheless, the structuralist agenda is particularly open to such applications.[91] Yet while literary texts can depend upon a fairly stable consensus with regard to vocabulary, grammar, rhetoric and poetic, selective databases need to address the problem of longevity more urgently.

One development that has been pursued by bibliographers is the move-ment into relational databases,[92] which require far less rigid selection of significant information, and which by consequence make possible a wider range of questions. A relational database structure is set up to allow for the definition of categories by specifying combinations of types of information relevant to any particular query. For example, in one query it might be helpful to define an author as a writer who has published in printed form, whereas for another purpose one might want to define an author simply as someone who makes verbal texts. Or it might be helpful to recast the grammatical categories for a textual analysis into rhetorical terms. From one point of view, relational databases within their own structure do some of the categorising work that bibliographers used to carry out as a preliminary to making their indexes. Indeed, they used to work far more slowly in computer terms just because of the additional organising they do. But in effect they can also implement the ideologically acceptable modes of organising that are dependent on subconsious or highly learned selectivity – such as grammar, punctuation, spelling and so on – that characterise databases created from texts whose verbal continuity has not been changed. To an extent, this is what Burrows is making the most of when he looks for modal verbs in Austen's novels and does so by feeding in a list of their variants.[93] Rather than adding specific mark-up codes to a text, he is counting on pre-existing grammatical conventions to indicate matches for a trained computer search. But the longevity of such a structure, because it is ensured by the way it implicitly carries cultural and social assumptions, is even more ambivalent in terms of the way it makes knowledge possible.

As with all information technology, the main problem is reminding ourselves to have good reasons for trusting the computerised organisation, because it is so easy to become dependent upon its systematic procedures. There are also different ways of responding to its procedures which once again throw forward the implications of methodology in the humanities. Take a selective database such as the one described above concerned with domestic texts: it already omits a great deal, hence when inquiring about the 'average' price of a book, it is necessary to remember that this is only probable and will not account for variations due to different binding practices for different audiences. Already the database has reduced the amount of description possible, even if it does so for the very good reason that the computer could not otherwise have coped with it. Reduction of possible observation by focusing on currently significant elements may mean more immediately available pieces of information, but also implies long-term reduction of knowledge.[94] The manner of reduction also means that the questions that can be asked have to restrict themselves to the system that is offered: again, short-term benefits but long-term limitations if we take the manner of reduction for granted. The kind of analysis offered by a text with little reduction, or little mark-up, is certainly limited by the lexical, grammatical and rhetorical structures of socially accepted devices for communication; but for many arts users who do this kind of study anyway, these limitations are more acceptable. Indeed, the limitations are often unexamined and slow to change. Using the computerised descriptions to guide one back to the text itself, rather than provide specific answers within the system of significance the descriptions offer, is a very different use of database organisation.

The gesture from artificial intelligence toward the arts

The gesture AI makes toward the arts as 'heroic' and 'beautiful', or toward their understanding and provision of metaphor and analogy, is dependent on notions of language as adequate to representation. Implicitly there is a hope in this gesture that arts strategies can help to construct a multitude of isolated and internally consistent pluralities. But the flexibility of analogy is ambivalent. It may provide parallel ways of knowing, but it may simply mislead because of its apparent points of comparability. It may be functioning entirely within a larger system, possibly that already defined by a set of assumptions laid down by a predicate calculus. Flexibility and comparability in themselves are no guarantee of social and historical appropriateness. They may shift a paradigmatic set toward knowing the real, yet provide no alternative to it: they may simply move the perspective. Indeed, analogy may not 'hear' that an alternative is necessary. Questions of heroism and beauty are often deaf to value.

If observations become data once we start questioning them because we impose a structure through our questions, then data becomes the key to the concept of significance. Significance is partly to do with the recognition of patterns and structures that our questions both desire and define. Two primary procedures for that recognition are hypothesis, which deals with the testable, and story or narrative, which deals with the contextual. Broadly speaking, these two procedures outline the approaches of, respectively, the sciences and the humanities. Hypothesis is based on a logic that aims toward the validating of possible worlds that are consistent in their own terms. On the other hand, narrative is often structured through topical reasoning that enacts the perceived interactions of current contexts and is frequently inconsistent. In coming to terms with computer use in the humanities, we need to understand both that hypothesis is an unreflective version of analogy, and that such selective masking of context can be helpful in its own ways. But it is also important to consider structures that can be responsive to the topical reasoning of narrative.

Any of the degrees of organisation of data, from the verbal texts implicitly structured by reading and writing skills, to the computerised texts with more or less mark-up indicating significance, to the selective texts of bibliographic analysis, can be isolated within ideological givens such as grammar just as much as mathematics is isolated within its rule-bound world. Just as AI finds it difficult to provide context for scientific inquiry, so computer applications in the humanities found it difficult to provide context for inquiry into the arts. Yet AI at least self-consciously foregrounds the difficulty. Computing in the humanities has tended not to do so. Indeed, having structured the data, there is a cruel inexorability about the path into numerical analysis and hypothesis testing that many computer users in the humanities then take. Literary and language activities do not describe phenomena in the same way that say 'gravity' describes phenomena; they are not 'testable' in the same way. But with the sheer number of detailed observations generated by questions to computerised databases there is a profound temptation to make them 'testable' by turning them into numerically assessed events; and the main tool that is used is statistics.

Historically, the use of statistics begins at exactly the moment in the late Renaissance that rhetorical reasoning to probable conclusions lost acceptibility;[95] statistics in its own way provides an alternative means of dealing with probability. Also for historical-cultural reasons, statistics can both provide indications of probability by way of tables of numerical significance relevant to particular mathematical calculations, and provide indications of probability through the visual clues yielded by histograms, graphs and various other visual mappings of number.[96] In the former method, our understanding of significance depends upon a table telling us that a particular number is significant: curiously exciting but unnervingly unreflexive. In the latter method, we bring at least some learned comprehension of visual significance to the picture the statistics present. At some stage we have learned that points falling on a straight line indicate some analogous property, or that the high mid-point of a graceful belled curve is an apogee of something. We are also, in our culture, relatively sophisticated at reading maps, so that when A. Moreton maps collocations in Shakespeare authorship studies on top of each other while T. Merriam[97] maps them along a binomial distribution in discretely presented visual shapes, we think the Merriam more persuasive, more convincing. Somehow, when Burrows produces through eigenvalue calculations individual 'voiceprints' for a writer which shift chronologically but appear to retain unique shape, we believe in this while we simultaneously lecture to our students on the disappearance of the essential identity of the author.[98] It is likely that individuals bring to visual significance their insistent exposure to the cultural use of geometric forms that has also characterised our culture since the Renaissance.

There is, as with all modes of ordering, nothing intrinsically wrong with statistics. Its organisation, like that of the bibliographic database, is simply further removed from the text and hence omits more than a computerised edition. If you like, statistics itself is a kind of second- or third-order database. But two points should be made: if humanities scholars or students are to proceed more often into the possibilities for numerical analysis offered by the sheer quantity and detail of computerised databases, then both those using these techniques and those attempting to assess them should understand far more about them. Not much attention has been paid to what the literary or linguistic critic does with either numerically or visually indicated significance. Contributions to the discussion that took place on these issues during the 1970s in the social sciences have emphasised the difference between exploratory and confirmatory statistics.[99] Exploratory statistics are those which guide you to a number of potential hypotheses, while confirmatory statistics test these hypotheses. The distinction is helpful, but it is unfortunate that the usual scientific environment for 'hypothesis' uses it to establish testable validity. Just so, there is a tendency for humanities users to employ statistics to imply that their results are in a sense 'true' because testable, and for those who do not use statistics to reject them as spurious because they eliminate and omit so much context. At the moment it would suggest that the responsibility lies with those who use them to indicate the methodology and hence provide the context, if they want their studies to be helpful.

The reason that numbers are so powerful is that they arise out of a set of definable human premises or rules or axioms, and as long as they stay within these they can give 'true' answers. But people and language do not work like that. Numbers often find it helpful to forget what is omitted, but language and literature fail if they do so. The problem many humanities users fear, and which is evidently present in many of the responses to databases and computers in the study of language and literature, is to do with turning language and literature into reductive fact and forgetting not only the conceptual procedures of the process but also the social context. There is an enormous amount of philosophical and epistemological work to be done in the area of what 'data' is for the arts, and what kind of knowledge people involved in computing for the arts are pursuing. Possibly, if the answer to this becomes clearer, then the gesture that artificial intelligence makes to the arts, which is a gesture toward the contact that science makes with the real, may find more substantial ground.

Artificial intelligence: rhetorical implications

Anyone engaged in studying verbal texts also knows that the kind of question asked is immediate to the individual: we all ask differently, even if within ideological parameters. Recognising a pattern implies remaining open to gatherings, groupings, clusters and repetitions, and responding to the internal and external relations they set up. However, this work is carried out with greater and lesser emphasis on outlining the methodology supported by theoretical consensus. In many cases there has been a profound lack of self-reflexivity in many discourses of the humanities, and recognition of this has led to the upsurge in courses on critical theory and cultural studies.

Concurrently, developments in AI have interesting parallels with developments in the humanities and in cultural studies. These latter have also moved from formalities into structuralisms, recognising the inadequacies of both. Yet the humanities are based on the belief that exact communication is impossible, that knowledge representation can never be accurate. Their response to the contextual inadequacies of structure and the formal symbolic, has been to develop a 'post-structuralism' that also looks for patterns in a limited context yet defines them as significant because they are only part of the story. Just so, deconstruction attempts to approach context by looking for significance in what is missing, omitted or not appropriately represented. Either of these approaches can have at least two rhetorical stances: the first takes the (in)adequate definition, the what-is-missing or differ*a*nt, as the basis for a game; the second takes these and foregrounds them, both to indicate the limitations of merely engaging in a systematic game and to insist upon the contingency, the pressing upon us, of material context.[100]

The reason that these are stances rather than strategies alone is that while the rhetor/writer is either overtly or covertly using the techniques and devices historically specific to a particular strategy, the audience's interaction with the text brings its own strategies to bear upon it. If the writer goes out of the way to foreground inadequacy yet the audience decides to respond systematically, then perspective will be lost. Stance refers to the rhetoric of the text inclusive of both rhetor and audience. AI at the moment appears to be working within the stance of systematic or designed games which, as with all games, is end-directed and based on hiding its rhetoric in order to control its audience. I suspect that this is so because whereas the humanities know and work with the limitations of representation, AI seems to desire its perfectibility. When it does not find such exactitude, as with 'natural' language, it simply moves on, does not address the difficulty, sees it as failure rather than as the potential location for context.[101]

However, AI needs to develop rhetorical awareness, particularly since it is contributing not only to the questioning of modern science but also to its extension and the technology that results. AI can and is being used to underwrite the validity of the plurality of possible worlds necessary to the economic functioning of modern technology. Technology needs possible worlds to establish the order, the grounds, that will allow it to fulfil its promises, realise the local desires that people have paid for. These possible worlds are the basis of the 'service' society that provides the satisfaction of a specialised need,[102] and they lie at the centre of so-called post-industrialism. Their plurality is necessary not only so that many people can profit, providing the illusion of liberty, but also because a large number of small worlds have a better chance of successfully gratifying specific desires: the bigger the possible world, the weaker its control over realisation, the greater the risk of intrusion by reality/context, and of failure. These worlds appear to grant freedoms and liberties: but they cover up the broader picture, they hide the industrial base in a plurality of commodity.

AI has developed a sophisticated understanding of this process. It now needs to cease looking at the limitations of knowledge representation as failure and instead provide an assessment of its techniques and strategies,[103] so that we can begin to evaluate the status of this plurality that the economics of modern technology is moving us into. It needs to become aware of its stance, to take a stand on it, and as a metacritical discourse it is better placed than many to do so.

Story: continued

Each scientific discipline, physics or chemistry or biology or any other, seems to have different etiquette. Yet because the physical world does not directly speak back, only through untranslatable resistances and compliances, the etiquette becomes a power-relation in which the scientist has to depend upon themselves to respect or otherwise deal with the respondent. Some people, rarely I suspect for I never practised it myself and only saw it once or twice, employ respect toward the natural in the same way that they respect other people. They understand the textuality of science as a practice that offers them a medium through which to contact other human beings. Some people engage in a more traditional aesthetics of science, playing with the social patterns of behaviour constructed by others, engaging with those human structures and speaking to and competing with the artefacts, rather than the natural world. And some people play games.

While learning the different etiquettes, in another stationary moment of memory, the young woman experienced the intense aesthetic joy of science. An organic chemistry teacher had set the second-year undergraduate class a problem that had brought the postgraduates to a halt. Coming back on the train on a Sunday night, the young woman pored over the question for whose resolution they had been given just three analytical tools. In her exhaustion the analytical drifted into the analogical and suddenly the resolution came to her. In passionate embrace the movements of longing and binding and rejection that colour the tropes for molecules and atoms, entered a courtly dance of taut/taught integrity, flooding her body with adrenalin and extenuating out and along the multitude of filamented patterns in every part of her physiology. When I re-member the ecstasy, it far outweighs the intensity of any early sexual experience.

At about that time there emerges another stasis, set in a terrain of war-torn lab benches, stained, littered with bits of metal and carefully crafted glassware looking ominous/lying anyhow, deprived of specific purpose. The ecstasy of the moment of resolution had led the young woman to think of science as experience that demanded poetry, even war poetry. One of the many craft skills that the scientist in training learns is how to write about the trials of the quest. This is partly to test the individual adequacy of each person, who may otherwise be working in a group, but it is also to ensure that each scientist can display the results of their work in the recognised medium for communication of legitimate knowledge: the written word. Technicians are not taught this literacy to the same extent; it is part of their world to remain silent or silenced in the private space of the laboratory, so that the scientist can voice knowledge. But the scientist has to acquire quite specific rhetorical strategies for doing so, and trains for this alongside the experimental work.

When the young woman tried to structure her reports with poetic device she transgressed centuries of agreements about the communication of scientific knowledge. Afterwards she found out about the diaries of Poincaré and Einstein, long discussions over naming and metaphor in Priestly and earlier, in Boyle, and the fantastic cartography of medieval alchemy and astrology which embroiders ecological understanding with narratives of specific geography. She realised that her mistake was not the use of metaphor but the use of metaphor as such and in public. There among the benches, in the trenches, she was instructed in the language of plain words and passive tension, by a graduate student tutor who quite liked metaphor and could not explain why it should not appear, only that it was necessary to erase the intensity, efface the emotion and restrain the passion of science.

But another etiquette also came to bear, marginally, on learing how to be a scientist. The teacher of physical chemistry once introduced to the class a scientist from Ghana. This man had spent several years after his postgraduate degree with his tribal community, working with the medicine-giver, documenting his cures, salves and preventive medicines. He had with him on that day a wooden (not metal) box full of small bottles, some with emblematic names like 'Cleopatra' for a gonorrhoea cure. Each bottle held the active ingredient for the appropriate cure for a particular disease or illness. Some he knew about and others he did not. But for all of them he was searching for a sponsor to isolate and identify the ingredient – a capital-intensive process that even then in the late 1960s/early 1970s, could bring nothing or could bring extraordinary financial reward. Nothing was said, nothing changed. But the young woman became confused: Was this a good thing, this exchange of medicines? Was it a rip-off? (And later) would the medicines work anywhere else? Was it an exploitation of the racially exotic? Or a helpful interrelation in a global community?

From this static panel of memory onward, the ethical and moral begin to intervene in the landscape to give it location and time. The lab technician, now trained for six years in craft, begins to get layered on top of the learning scientist and the rather older young woman. It seems, in retrospect, appalling that someone could have got as far as the final year of a university degree course before beginning to think through the social effects of the scientific knowledge that was being learned. It is no less unnerving that the lab technician took six years to think about it, but it is more understandable since the lower in the hierarchy one works, the more effort and time have to be put into simple survival. As the young woman stood back, began to work as a technician but with the more self-conscious knowledge of the scientist, the romance structure of the social world of science became clearer at the same time as broader social implications of that world began to seep through the membrane, staining the filaments with organic material, corporeal and corruptible, capable of genesis, decay and inexplicable synergy.

The newly self-conscious awareness lent stereotypical features to the structure of harassment during one job for a commercial scientist renting a bench in the university graduate lab. As the experiments were timed for later and later at night, and after one assault, the young woman stopped working for him. Four days later the graduate tutor who had instructed her in plain language rang up to find out whether she was unwell, anxious because she had apparently been working with chemicals known to have potential for degeneration of the nervous system. Interesting that her employer had not told anyone she had left; interesting that the graduate community cared enough to check her out, although not enough to warn her in the first place; and interesting that of course she did not say why she had left.

The methodologies of her work had taken her into many labs where animals were used, but as the social implications of the science stained the glass, so they let in different and unfocused light on these practices. Shortly after leaving science for the humanities, she took a job to support her arts habit with a medical lab doing work on spina bifida babies. The cruelly deformed foetuses of this brutal tragedy were at least not intentional grotesques, but the pain of working on their behalf recast the corpses and mutilations of her other work in anguished burdens of reflection. Moving still closer to human beings, her final lab work was in a hospital where the quest structure of research science was muted in response to all those immediate needs demanding their own trials and resolutions.

While still inside the glass womb, she began to revel in the subversive cells of female company into which the lab technician work pulled her. Suddenly so plain: she had never been taught by a woman, she had never been hired by a woman, but there were all these women all around her inside the glass, working sometimes as research assistants but more often like herself as technicians. And what they all enjoyed in amongst the tedium of repetition was the practice, the careful measurements, the skilled manoeuvres with hands, bodies, mouths, noses, eyes, ears: the whole body a sensory respondent. She saw that these bodies passed easily back and forth through the glass membrane, sometimes dragging their filaments like sad entrails behind them, but sometimes first packing them up and neatly putting them away or moving them wholesale into the other world. She learned from them that it was possible to leave the epic drama of the quest, not only for the coffee room but for the other world. She not only saw the stain on the glass left by every social infringement, but recognised that the other world could get inside the glass, that it had been able to do so all along. And that the figures in the filamented landscape could speak, if they chose, to those outside, those in Lydia's different life.

Moving closer to human beings in an obvious and concrete way changes the understanding of what science can do. It may simply be a game with sets of rules that the natural world does not resist; it may allow its own aesthetic pleasures that recall the pleasure of a kind of recognised textuality as people contact the physical environment around them. Yet the moral activity in working with science, and its ethical actions and effects, make it clear that it is at its centre mainly another way for human beings to communicate with other human beings: science is a text through which we act on, engage with and affect other people. But I did not recognise this until I had learned the same thing about the arts and humanities.

3 AI and representation

A study of a rhetorical context for intellectual legitimacy

In the short history of AI it is evident that it has begun to explore a variety of techniques and methods, but it has yet to learn how to involve a sense of positioned ethos in order to offer contexts for them: this critical strategy is often referred to as reflexivity. And although reflexiveness is a step along the way to considered methodology and processes of assessing social value, in rhetorical terms AI needs not only an understanding of ethos and pathos, the participant and the responsive context, but also an awareness of stance which provides guidelines for halting continual reflexivity in order to make decisions about necessary social action. Mainstream philosophy of science, which often turns to computing science and AI because of their claims to display the representation of the natural sciences, offers an analysis implicitly founded on the understanding that language is (in)adequate. This chapter looks at a number of analyses, particularly those which use science as part of a larger picture of political philosophy, and comments on the split perspective that the understanding generates: that representation can be either only a coded reference to the world or a strategy that attempts to transgress and transcend that code.

In 1991 a lucid and empowering essay by David Kirsh outlined a variety of approaches to AI, and took on the big issues of cognition, representation and learning. In the process of doing so he positioned as a central concern questions to do with how we come to agree to common grounds for representation, and charted the movement by some AI researchers from axiom to plural perspectives. While the critique of logicist assumptions is welcome and in these hands sophisticated, the plurality that results does not address issues of social immediacy and action: it remains with the contexts of ethos and does not take on the practicalities of stance. More and more commentators recognise the need for guidelines to social interaction in their talk about evaluation: how do we know when AI can be said to be valuable? More to the point, how do we know that value is not just success in fulfilling the internally constituted rules of self-defining worlds, but is rather an intersection with social needs that offers ways of addressing them? Evaluation is pertinent to all areas of inquiry, but particularly to AI at the moment, because there are so few rules of thumb for the way in which it deals with representation and touches upon reality.

Possibly because of its intimate relationship with modern science and its attendant stress on method, AI has concentrated on finding techniques and plausible common grounds for its work. In rhetoric, plausible argument works from opinion, and argues from basic assumptions held by a group of people. In contrast, probable argument first attempts to discuss commonly held ground, decides upon its appropriateness for the argument in question, and only then proceeds to argue from it. There has been a lot of commentary on the advantages and disadvantages of plausible and/or probable argument, with most writers agreeing that the plausible is the root of deception and manipulation. There is also a line that rejects rhetoric entirely and claims that since truth about certain things is absolute and can be known and articulated, basic assumptions do not need to be discussed: in other words the 'plausible' is not relevant to 'certain' things, defined of course by the individual and/or hegemonic group on behalf of others. Rhetoricians frequently concur that probable argument is the more testing and helpful, but that plausible argument is often necessary because of time constraints or the need for immediate action.

At least since Aristotle, science and more generally ordered knowledge has been described as an activity which need not engage in probable argument all the time. One of the reasons given for this is that within a field of knowledge people who are part of that field by definition have worked through the basic assumptions. They are pursuing that science only because they have entered a club, a culture with ground to which they have given prior assent. The problem, of course, is that the activity of assent is soon forgotten for a variety of reasons[1] and the door is closed on the reassessments enabled by probable argument. Indeed, science has come to be known in broad terms as the field of intellectual pursuit in which people behave as if they are working with 'certainties', or at the very least as if there are basic assumptions that do not need to be questioned, that are independent of context and appropriateness.

Kirsh defines AI as based on the description of a consensual reality with adequate vocabulary for representing concepts that underpin 'the millions of things we all know and that we assume everyone else knows'.[2] But having been represented, the description is rarely questioned. It is perhaps significant that the fields of inquiry recognised as impinging most directly on AI, from philosophy and linguistics to theoretical computer science and cognitive science, are not those demanding social immediacy. On the other hand, inquiry in the humanities is seen to have retained an abiding interest in precisely what makes for appropriateness and attention to social context.

If AI is to assess appropriateness, it has a specific problem because it is still perceived as more relevant to the sciences than the arts and humanities. Assessment of the common grounds for representation will have to push open the door to rhetoric that Aristotle pushed toward closing, and will have to move the public debates of science from the plausible to the probable. The plausible can be acceptable and even fun, for worlds of games that do not impinge on social relations of dominance and power. But where the social systems of scientific games becomes increasingly influential through various technologies and engineerings, it enters a world of political action that makes assessment essential. The first stage of assessment is to question, both what is appropriate to say and what is necessary to say but difficult.

RHETORIC, REFLEXIVITY AND STANCE

A rhetorical approach to the politics of representation that results when the arts and humanities come to deal with computers and computing,[3] sets up a discussion about the illusions of power offered by closed systems, and the problem of knowing when they have ceased to be helpful or whether it is even possible to open them up. This discussion is analogous to the much broader debate about the political implications of communication that has been going on in Western European political philosophy for some time, as it has attempted to cope with the effects of the sudden authorisation in the early twentieth century of new voices from the recently enfranchised adult population. Suddenly the whole process of how one finds and assesses common grounds, which had been worked out even in the post-medieval world in terms of effective power politics for a small elite, was placed in the world of a widely varied and huge public.

The distinction made between the probable and the plausible, outlined in Chapter 2, was a distinction intended to help people faced with many different versions of events and possible ways of acting, to decide upon the most appropriate grounds from which to work: grounds appropriate to the 'good' of most of the citizens. Aristotle was working in an historical context of many powerfully persuasive voices, the sophists, who were often from outside Athens and brought to it many different backgrounds and needs. Even strangers, if they were men, could apply for citizenship of Athens; although women, many of whom are also documented as having powerfully persuasive voices, could not. Both Plato and later Aristotle write in various texts that indicate their understanding of and respect for sophistic rhetoric which challenged fixity, stability and authority, from a variety of perspectives. But they were also, Aristotle more than Plato, concerned that multiple voices and perspectives can end either in a kind of political free-for-all where might is right, or in a muddle.[4] The question became: given the activity of many positions, how does one agree to common grounds that enable us to make decisions and take action?

The plausible, which is based on self-justified opinion, is simply not a sensible or ethical way to proceed. The probable, with its strategies of negotiation and argument, is far more engaged with the complex immediate contingencies and far more likely to arrive at grounds appropriate to the larger number of people. However, the possibility of differentiating between the necessary and immediate agreement about appropriate common grounds (the probable) and habitual agreement about them (the plausible) has been problematised by the sheer size and number of many divergent and conflicting needs and desires. Both consensus argumentation directed to a decision about action in which all the discussants participate,[5] and corporate argumentation in favour of an action which will be taken on behalf of the participants,[6] may lead to tacit agreement about adequate representation. The latter shift from corporate to totalitarian argument has been well documented,[7] but the elision from consensus to the self-enclosed cynicism and nostalgia of the pluralist worlds of hyperliberal club culture has only recently begun to be described.[8]

The activity of modern science has much in common with club culture but is also working, significantly, in a representative mode that is completely different to that of the humanities. Furthermore, it is not surprising that science has become a central motif in the discussions about a 'legitimation crisis' in the state, given that science is the one field where working with plausible argument or opinion, as if it were certain, is expected and authorised by the representative mode. Current descriptions refer to science working within the self-enclosed world of 'paradigms', either large-scale Kuhnian structures or the competing pluralities of postmodernism described by Lyotard.[9] If writers such as Gadamer and Habermas in the 1970s and 1980s proposed, respectively, a return to probable argument through assessed consensus discussion, or a return to corporate argumentation in universals,[10] others, such as Rorty following a line often incorrectly deduced from Althusser, argue that the distinction between the probable and the plausible is not relevant because we all do live in enclosed club cultures, therefore the probable is always opinion: there is no need to distinguish between an appropriate and an inappropriate ground, there is no need to assess social context.

My position here is this: that it is necessary to assess social context, to discuss appropriateness, to attempt to distinguish the probable from the plausible. This chapter will argue that AI has been trapped, for specific reasons, into duplicating club culture when it could be precisely the place where science attempts a different representative mode.[11]

REFLEXIVITY AND HOW TO STOP IT

If AI is to take a stance on the activities of science and technology that it appears to describe, it needs first to become self-conscious about its own techniques and its strategies. AI is about finding ways to represent the activities of human intelligence: both how we represent the world to ourselves (knowledge representation) and how we deal with that representation, respond to it, interact with it (expert systems), use it, and so on.

Representation is a complex area, and it is not the intention here to limit it to even sophisticated versions of visual verisimilitude or structural congruence. Rather, I would like to place AI within the carefully detailed debates to which the long history of representation in the arts and humanities has led. This is to underline the rhetorical work of representing which involves not only *technique* and device, but discussion of *strategy*, and most importantly an assessment of *stance*. It is within stance that the dilemma about evaluation in humanities computing can be addressed. And it is questions about evaluation in AI that need urgently to be answered before it can sensibly position itself with regard to technology and power. This discussion will focus on issues of self-consciousness, strategy and reflexivity.

A consideration of strategy has to include a sense of where the speaker or author or writer is placed within the description. In traditional rhetoric, 'ethos' implied taking up a particular role or perspective with relation to the audience, whose response was in turn cast as 'pathos'. Ethos implies a fairly stable set of conventions about the relation between the speaker and the discussion; for example, written genres like the letter or report, assume certain strategic relations once they are realised; we do not usually expect a report to be a lyric poem. Within current rhetorical studies generally there is growing concern with an increasingly far more varied audience[12] and subtly elaborated questions about the effect of the speaker on the discussion. In literary 'deconstruction' and the 'reflexivity' of the social sciences, there has been an extension of ethos to include an on-going assessment of how participation affects/effects the event.[13]

If we look reflexively or self-consciously at how our participation in both an event and its representation affects that event or representation,[14] questions are rather obviously raised about various parts of our knowledge and perception that are not adequately represented. From difficult questions such as 'why do you feel depressed?', to apparently simple ones such as 'why do you like cheese?', we often find words and other media inadequate to the representation of what we feel or know. Reflexivity is on-going and continuous precisely because that knowledge cannot satisfactorily be communicated, and this indicates a further problem: reflexivity is a useful tool but its continuous activity does not encourage positions to be taken from which decisions can be made and specific actions carried out.

The problem is how and where to halt reflexivity. On-going questioning is frequently halted at points where representation appears inadequate. We stop and attempt some kind of communicative context for a practice that resists discussion, but materially indicates some kind of knowledge. But if this attempt to voice tacit knowledge is made merely in response to an inadequacy, then reflexivity simply starts again and moves on to the next (arbitrary) difficulty. Used in this manner, the tool becomes intensely subjective. It works within a system of absolute/arbitrary dichotomies which define much of the social action that can be taken by the individual as a subject, and science is an intensely ideological activity, operating as it does within a discrete, institutionally sanctioned representation.

Communication and its pair, 'tacit' knowledge, lies at the heart of the debates about legitimation in science: is tacit knowledge uncommunicated because it is impossible to represent, because people have not yet tried to represent it, because it has become habit, because it is blind prejudice, or because it is 'true' and does not need communication? This range of questions sits parallel with the range of AI approaches listed by Kirsh: from distributed AI to moboticists to connectionists to Soar to logicists.[15] Without social context it is impossible to distinguish among the approaches on issues of knowledge and textuality. With an on-going use of reflexivity there is no sense of assessing which of many difficulties it is necessary to deal with, nor is there any sense of placing an attempt at representation before a public, for purposes of evaluation (how appropriate is it in addressing social needs?) or critique (is it appropriate at all?). This is why strategy, whether stable ethos or on-going reflexivity and deconstruction, needs stance that places it within social contingencies.

This version of the 'legitimation crisis' which stresses the problem of distinguishing the habitual from the appropriate public and social common grounds for representation, is related to the problem in cognitivist rational analytics of the distinction between prejudice, which is accepted blindly without question, and the axiom or explicitly stated rule.[16] Yet the two versions are not entirely congruent because the cognitivist dilemma does not deal with the difference or distinction between regulative and constitutive rules, rule-following and rule-performing, that makes the debate about habitual/appropriate analogous to the further problem of how to distinguish self-evident from working knowledge[17] or tacit knowledge,[18] and indeed how to distinguish these two, habitual and appropriate, from each other, or within each one among different strategies. These issues are directly relevant to theorising about evaluation in AI, which has moved inexorably toward concepts of practical knowledge, work and performance. This has happened largely because the aims of AI to represent knowledge and to represent expertise, cannot be satisfied by equivalating knowledge to information, or expertise to regulative rule-following. But in order to understand the necessary concepts, AI must learn how to become self-conscious about its techniques and methods.

Evaluation for the arts and humanities in the post-medieval period has largely focused on reflexive questioning of common grounds or topics as they are conveyed by representations within the ideology-subject axis, even though much humanistic activity does not do this. Philosophy, art, and what is temporarily called theory, has become the recognised place where we attempt to find new common grounds and try out their appropriateness, as well as where we assess the continued appropriateness of others. Theory is not prescriptive, but constructive and descriptive. The whole point of theory is that it addresses the areas of knowledge or perception in our lives that we find difficult to talk about (phronesis)[19] and brings them in contact with the craft of communicating (techne), so that we can try to work out articulations adequate to the context of our day-to-day lives, that will eventually enable us to make decisions about events and to act. A complexity that will be pursued at length in Chapter 6 is that philosophy, art and theory still largely work within the ideology-subject axis and hence like the commentators on science and artificial intelligence discussed below, take the tacit either to refresh the habitual or reveal the ideologically hidden. Too often they offer 'context' simply as selected perspective[20] rather than dealing with socio-historical immediacies.

Appropriate action is the broader significance of the word 'proper' as in: searching for a 'proper meaning'.[21] In the humanities this activity is sharply noticeable in the theoretical debates about gender and race, in which people who work within a dominant discourse, a dominant set of common grounds and strategies, that ignores and even represses their different knowledge, have shifted away from opposing the repression to attempting to find ways of articulating that repressed knowledge. The debates about 'class' are particularly difficult to recast in terms not of opposition but of unarticulated knowledge. Each area needs to take certain norms of communication for granted in its theory, or literally it would appear to be talking nonsense. Communication of what is known but not yet articulated is a process involving the speaker, the medium and the audience. If the audience has no *common* ground of communication at all, neither will it be about to make significance or sense of the text.[22] At the same time each area is self-consciously assessing common grounds and attempting new ones, and that reflexivity is not unitary because each one has different guidelines about where to halt and sort out the value of its representations.

Modern science differs from the arts and humanities in that rather than having other human beings or other human-created texts as its referent, it claims the natural world and the constructions that people have made out of it. This difference has resulted in a profound difference in the way that the sciences and humanities communicate to any public audience. For the 'hard' sciences, knowledge is about the natural world (phronesis); it is known from doing the experimentation (techne) and practising scientists are fairly continually reflexive about this engagement with reality by way of assessing the grounds of their scientific method.[23] But while the referent participates in this, it does not talk back. Furthermore, communication about this knowledge to a public is a second-order articulation about the first-order experiment, and therefore finds it easy to look for normative and conventional grounds for expression rather than reflexively assessing their appropriateness.[24] The knowledge is supposedly about the world, not about a changing communicative society; the communication is already at one remove from the real activity.

The attempt at methodology which has come to define modern science in the western world since the seventeenth century, can be seen as an attempt to provide precisely those guidelines that halt the reflexivity that public discussion entails.[25] But this attempt to halt reflexivity is not in aid of evaluating the grounds appropriate to social action. In science, the stress on methodology and on techniques of experimentation, can be seen as an attempt to question the unsatisfactory area of the plausible grounds of scientific argumentation, by taking science *out* of the social demands of communication. The desire of the Royal Society for a pure cognitivist language can be recast, not in terms of a concern for the objectivity of language for its own sake so that it can refer to the *world* precisely and accurately, but for the objectivity of language so that it can represent the *experiment* exactly. Furthermore the new science needed a plain language without its own textuality because like many Protestant reformers of that early time, it wanted and still often wants to deal with a natural-world reality that can fill out or complete the emptiness of words as they stand by themselves, as if words were bags in which to put the reality known in the experiment. This denial of textuality is exacerbated from another direction as technology begins to use science to make commercial profit from the eighteenth into the nineteenth century. What permits technology to commodify or make money out of science is its ability to represent science mechanically[26] as a fixed methodology producing 'true' products.

It is a self-defeating agenda: science defines language as inadequate to the representation of the experience of reality, and suitable only for the reporting of methodology. Hence the public understanding of science is restricted to the now clichéd version of scientific prose as a parody of technical rhetorical structure: propositio, inventio, narratio, conclusio; and scientific language as rational, analytical and cognitivist.[27] By closing the door on to society, in an attempt to get rid of the merely plausible, science only succeeded in restricting to physical demonstration its communication about what it was effectively concerned with,[28] and in producing a set of representations of its methodology that did not take account of common grounds – which is exactly where the plausible comes from: unassessed habitual agreement.

One of the commentators to have emerged from the debate over the legitimation crisis, whose work is pertinent because it appears to applaud and authorise the plausible, as community habits of interaction, is R. Rorty. Rorty rejects textuality on the grounds that representation is inadequately referential,[29] and in this duplicates the rejection of rhetorical stance enacted by science when it accepts that written language cannot convey the action of experimentation. He notes that this inadequacy of representation raises insoluble questions about the absolute and the relative, and wishes to replace it with 'habits of acting' and interaction with a 'neighbourhood'.[30] It is the case that if a stance is adopted that wants man-made control over all aspects of language that model it as a mirror of the world, then these dichotomies do emerge; indeed most of twentieth-century psychoanalysis engages with this and many other consequences of the stance of fantasy which technically this is.[31] But there are other choices of rhetorical stance such as allegory or dialogism, or the variety of poetic stances that have been built by people who have agreed to look at the appropriateness of the structures of representation rather than dismiss the problems that result from belatedly recognising that language is not exact.

What is curious about Rorty is that instead of choosing another stance, he stays with this impoverished notion of representation, and because of its impoverishment chooses to ignore it altogether. His arguments suggest that if representation is dropped, then metaphysics and epistemology can be replaced by politics: which is to assume that metaphysics and episte-mology have never had anything to do with politics, whereas if rhetorical stance and social context is retained then so is the political dimension. In effect, in dropping representation, analysis leading to political action is evaded; for Rorty, phronesis never needs to be articulated – indeed articulating or textualising is seen as a reductive and ultimately intolerant act.[32] However, as the rejection of rhetoric by science underlines,[33] dropping representation and avoiding textuality simply disguises the fact that all communication involves power-relations. Discourses that pretend otherwise, like science, become 'the most potent instrument(s)' of power for persuasion in our society.[34] The reason that Rorty's position is curious is that the comfortable elision into competency that marks exclusive club cultures and their excesses, from science/technology to liberal/authoritarian regimes,[35] is something those societies frequently attempt to resist yet which Rorty offers as a desirable end: and desirable is just what it is. Elsewhere he characterises 'habits of acting' as the private narcissism and public pragmatism[36] that make up the solidarity of club culture. It is, however, only the 'leisure' and 'civilisation' of bourgeois liberalism (25) that permits the desires of private narcissism to remain 'tolerant' rather than brutal.[37]

Bourgeois liberalism assumes a large measure of individual autonomy, but it is in effect dependent on an institutionalised state structure that creates conditions of private autonomy within a highly regulated public practice. Rorty suggests that 'civilised' culture is autonomous from the institutional regulations of police and bureaucrats (26), whereas it can only behave *as if* it were so by refusing to become aware of the institutional structure. His suggestion works without an understanding of the civic, that public space that is without active political assent, and can only consent tacitly to the structures of ruling power. Tacit consent underlines the need for the doublethink of simultaneously acknowledging state structure and then forgetting it in order to operate as a subject-pragmatist. Rorty's lack of awareness of this tacit, indicates how deeply embedded is his approach in the structures it seeks to evade, and which rather surprisingly aligns it with versions of Althusserian state apparatuses,[38] although it owes its primary allegiance to Feyerabend. The doublethink is here called 'ethnocentrism', and is contrasted in an article specifically on science, with rationalism and its 'criteria for success [which are] laid down in advance'.[39] But in effect ethnocentrism is simply what happens to rationalism when it forgets the need to look at common grounds, treats them as givens and does away with textuality. While the reflexivity of scientific experiment is its strength, its rejection of the textuality of communication has left it to build increasingly enclosed worlds, small circles of private knowledge exchanged by means of a specialist jargon: these worlds are often valuable but frequently become alarmingly narrow and reductive, and only continue to exist because of institutional support.[40]

In contrast, for the arts and humanities, communication in a medium is a first-order activity. Because they are concerned with human referents which engage in the activity of knowing, communication itself becomes the site of knowledge. Which is not to say that there is a necessary connection with society: the further away the arts move from public discourse the closer they come to duplicating, as in for example the nineteenth-century avant-garde, the small circles of private knowledge typical of the sciences.[41] But both the arts and the humanities find common grounds in the activity of their human texts, in the mimetic performing, repeating, imitating of these texts.[42] This is one of the reasons why it is so important for arts and humanities users of computing to understand the constraints of the system offered by a software text,[43] let alone hardware: they need to know not only what the rules of the software will permit, but how to perform its textuality within the large literary critical context. It is in this latter activity that they can reach evaluations, assessments of the knowledge that the interaction of software text and literary text has enabled.

Because the communication of their knowledge is part of that knowledge, the humanities have also traditionally claimed to focus on reflexive social questioning of methodology at the same time as using it, although they have not always carried it out. But when they do so, they help to maintain the socially immediate criteria by which we know how the common grounds of agreement and representation are being assessed, and to encourage attempts to be clear about the relations of convention and dominance that are involved, and about the relation between articulated and unarticulated knowledge. Such evaluation of stance helps to make clear whether we are seeking new common grounds, rationally extending old ones, breaking up existing common grounds, breaking through existing common grounds, offering different ones, replacing old ones, assessing old ones, using old ones self-evidently or fictionally or artificially, and so on.[44] However, as Chapter 6 will argue, the stance is most often from within the ideology-subject axis and the discourses of local interaction along its system.

AI AND THE REPRESENTATION OF SCIENTIFIC METHODOLOGY

For AI, both articulated and unarticulated knowledge are related to the activities that mimic human intelligence.[45] Yet because AI conventionally provides a theoretical space for working out the representation of science and technology, most AI currently deals in representation as if it were only cognitivist.[46] It is as if it were taking the limited representations of methodology in science and technology for the first-order actions (phronesis) necessary to repeat and imitate (techne) in order to articulate knowledge. Hence it is concerned not with a textuality which encourages repetition of an activity that locates (mimesis), finds and assesses the appropriateness of common grounds for all kinds of knowledge and perception, but with a stable set of common grounds that evades the implications of their plausible basis, and seeks only to represent experimentation, or at worst to test the 'success' or 'failure' of that representation. The frustration with this approach is becoming more evident in AI writing itself, with some writers claiming to work without representation at all because it is self-delusory, or to shift it into plurality.[47]

The problems here are specific: first, that if assessment is not self-conscious, the old common grounds will be taken as self-evident; yet second, that constant testing is an arbitrary gamesmanship that lets one win or lose but does not offer new common grounds for action. Since the communication of science and the commercial activity of technology[48] depends upon treating the self-evident as axiomatic and the unarticulated as irrelevant knowledge, AI is restricted to the process of looking for rationally acceptable topoi – models, digital analogues, frames[49] – in order to represent. Because of the restriction of cognitivism, indeed Hubert Dreyfus defines a 'cognitivist' as a rationalist with a computer;[50] AI tends to find these representations in regulative rule-bound systems.[51]

This tendency to reproduce the patterns of intensely private theory, and to work within the structures of artificially closed worlds, is what ties AI to fantasy, to the rhetorical stance of technology and industrial capitalism[52] which attempts to create isolated worlds that deny larger social and political interaction.[53] Yet it is precisely there that assessing the appropriateness of common grounds is not only most difficult but also necessary. The concepts most frequently discussed by people involved in theorising AI are intimately tied to the techniques for building self-defining worlds. The central devices are: algorithmic knowledge,[54] predicative knowledge,[55] digital representation, propositional descriptions,[56] regulative rules[57] and determinist structures. Each is part of a doublet that presents its other face, which is frequently offered as the reflexive alternative: mathematical knowledge, tacit knowledge, analogical representation, narrative description, constitutive rules and relativism. What this chapter will go on to do is discuss the usefulness of these doublets to AI in any attempt to contextualise its methods and become reflexive about its practices. The discussion here will argue not that the determinate and relativist doublet can be ignored, or that it should be evaded,[58] but that AI has failed to pursue reflexivity because it was misled into taking the communication of scientific method for the practice, and has as a result avoided the complexities of representation, particularly of constructing new grounds for unarticulated knowledge and perception. The immediate questions then become: how can representation become self-conscious or reflexive? What is unarticulated knowledge? What can it be for AI? Can it become articulated, and if so, how? How can we halt reflexivity in social appropriateness?

AI AND THE ARTICULATION OF WORKING OR TACIT KNOWLEDGE

As the limitations of rationalist, cognitivist approaches have become insistent, AI has crept toward the social by introducing contextual factors. The predetermined common grounds of the methodologies may satisfy attempts to know fact, information and data, but they fail to ask about knowledge of practice which demands not only specific context but also an understanding of stance.[59] There is a striking problem here: if practice, or working knowledge of how something is done, is often silent and inarticulable, how can AI represent it? One of the anxieties about working with models provided by AI derives from this awareness that because the models give the illusion of completeness, people working with them will lose the unarticulated practical knowledge that comes from working with the actual world.[60] But if the representations work by encouraging reflexivity about their limits and exclusions, then the user gets access to analogical ways of representing practical knowledge, that are at the same time ways of participating in that knowledge.[61] One result of this latter approach is the recognition that what people who work with computers need above all is training in representation,[62] not just in information representation but in a whole set of skills developed in the arts and humanities for representing a much wider field of knowledge.

DREYFUS AND JANIK

If AI has crept toward context, many theorists of philosophy and epistemology have explicitly raised social questions about AI and practical knowledge. So what is practical or tacit knowledge? The point of agreement among many differences is that tacit knowledge is knowledge by the expert, the knowledge of expertise that is learned in practice. In an attempt to illustrate why AI models often beat beginners but rarely measure up to experts, Dreyfus outlines a 'phenomenology of skilled behaviour' that moves from novice, to advanced beginner, to competence, to proficiency, to expertise. In the process he describes a gradual internalisation of heuristic procedures so that the proficient performer is able to understand a problem with immediacy and then 'must deliberate about how best'[63] to act; however, the expert has internalised both understanding and action, so that 'he [*sic*] does not solve *problems*. He does what in his experience has normally worked and, naturally, it normally works' (106). What is missing from this account is any attempt to describe how the 'intuitive' expertise differs from cognitive heuristics; nor does Dreyfus attempt a representation of expertise. The remainder of the illustration he offers ascribes the roots of cognitivism in western epistemology to Plato, which is ironic since Plato attempts one of the most enduring representations of non-cognitivist expertise in his accounts of writing, medicine and gardening in the *Phaedrus*.[64]

It is precisely this problem with the representation or articulation of expertise that is central to the future of AI. Richard Ennals has called it the 'bottleneck' to AI development; and Alan Janik, who has been a guiding hand to many commentators interested in the implications of practice for AI, makes the possibility of articulation the focal point in his definition of the tacit knowledge of expertise. Janik distinguishes between tacit knowledge that can be articulated but happens not to be, such as guild secrets or recipes,[65] and tacit knowledge which cannot be articulated even if we wanted to do so: here he offers the two examples of 'certain non-visual sensory experience and the procedure involved in following a specific rule' (55). The elucidation of both kinds of tacit knowledge for labour management, and for understanding the 'limits of the possible' in actual working contexts, is clear and helpful. It provides a way of talking about 'quality' within the economics of production (61), as well as the basis for a critique of both Habermas and Gadamer. Janik describes the irrelevancy of universalist/generalising principles of legitimation for labour relations working within structures of domination; but he also insists on a recognition of the limits of any particular community or set of actual conditions. This is essential if philosophers are to halt the continual reflexiveness typical of the cynicism and nostalgia of postmodernism.

However, there is a problem with the example of 'impossible to articulate' knowledge. The case of non-visual sensory experience that is offered is the 'smell of coffee' (56): because we do not have an articulation adequate to describing the smell so that someone unfamiliar with it could recognise it, we have to supply metaphors and analogies. What Janik does not go on to consider is what happens then.[66] In common with agreements about visual representation, there are frequent agreements made by particular groups of people about the adequacy of an articulation representing other sensory experience. People not part of a group will have as much problem recognising visual representation, as they will with representation of any other sensory experience. To this extent, the visual is not distinguishable from other senses, although it often appears to be so because technological developments in global media have educated a considerable part of the world's population to specific and sophisticated agreements about it. And, to the same extent as the visual, the other senses are open to education in representation. Wittgenstein uses the same example, the aroma of coffee, to discuss the way that we experience a missing description,[67] but he is far less definite about the im/possibility of finding a verbal articulation for it.

More interesting, but also more problematic, is Janik's description of how constitutive rule-following resists articulation, and this is an argument also informed by Wittgenstein. Janik notes:

> Rule following activity entails the kind of knowledge that is only ac-
> quired through repetition or practice…but terminates in creative activity
> as we learn that we have to guess how to continue to follow the example
> we have been given and, ultimately, as we learn to invent new ways of
> carrying on….In short it is an analogical rather than a digital activ-
> ity….Moreover, if constitutive rules were known before their applica-
> tion, we could never learn to apply rules without more rules.
>
> (Janik 1987: 578)

What is not considered here is the way that constitutive rules tend to become regulative with repetition, with time and with certain social contexts.[68] The way in which social context effectively brings together both regulative and constitutive, is underlined by Wittgenstein when he refers to the 'tacit presuppositions' on which any analogical 'language-game' rests.[69] Differ- ently, J. Austin directs the latter part of *How to Do Things with Words*[70] at distinguishing between what he calls the constative (descriptive) and the performative (constructive), as part of a larger attempt to distinguish between locutionary (correspondence with the facts) and illocutionary utterances. He notes that while the constative may need to be 'right' in the sense of being 'true' to 'your knowledge of the facts', both constative and performative may raise questions about 'right' in the sense of whether 'this was the proper thing to say' (145). Social context, or 'the total speech act in the total speech situation is the *only actual* phenomenon' (148) which may elucidate the distinction.

Social context is brought together with time and repetition in the consis- tent and coherent history of the process of topical or analogical reasoning documented by the arts and humanities.[71] A topos or a common ground will be hammered out into language among a number of people attempting to find a place, literally the ground upon which to discuss and decide issues. Janik himself uses the phrase 'seeing where we are' (1987: 57). Once the topos is 'in place' and has provided a helpful ground for learning about issues, the sense of its specific appropriateness elides into generalisation. The related procedure of the 'cliché', or 'cliquer' from the clicking of type into a printer's composing stick which allows for multiple repetitions of the same thing from the printing press, is exactly to repeat an appropriate analogy so often that it becomes regulative, self-evident and banal.[72] At the same time, repetition loses the foregrounded artificiality of the analogy or 'canonised example' or 'model' or 'pattern'[73] that is representing practical knowledge, and the analogy becomes an accepted convention for that knowledge.

The articulation of practice will always be under historical and social pressure; hence the articulations underpinning regulative rules will become inappropriate, and new common grounds, genres, canonical models and topoi will encourage and necessitate new articulations. Whether the 'need' is answered is often dependent on financial and economic strictures. Janik hints at the importance of history and of social necessity in his criticism of Hubert Dreyfus as a 'phenomenologist' who does not distinguish between constitutive and regulative rule-following. Here Janik notes that regulative rules are 'only possible on the basis of the former [constitutive] rules',[74] but like Dreyfus he does not expand on the historical process by which the one representation can lead to the other.

WITTGENSTEIN

Wittgenstein's *Philosophical Investigations* and much of his later writing is precisely concerned with a study of how we can represent 'what is not the case',[75] which may account for the considerable number of commentators on practical or tacit knowledge who turn to his work. Wittgenstein defines 'philosophy' much as the humanities define 'theory', as discovery and description (47, 48) of what happens when normal, prescriptive rules do not work, when we get 'tangled up' in the rules of the language game (50). Just as teaching and learning need to be by way of example and practice (63–4) rather than by propositions, so must language, in providing practical knowledge of reality, not limit itself to conceptual/cognitive representation (45–6). The language game is not an explanation but the *'primary* thing' (167). It must provide models as 'comparative' not dogmatic (51); it must understand the need for 'appropriate' words (54). Common grounds are not 'good' simply because they make events more likely (the plausible), but because they have an experiential 'influence on the event', they are part of it, they make it 'really probable' (136).

Underlying Wittgenstein's investigations is a careful critique of the practices of cognitive and behavioural psychology contemporaneous with his writing. Frequently, the writer distinguishes between human 'sciences' and natural sciences on the grounds that study of the former depends not upon measurement, calculation, instrumental concepts, but upon the external sensory reactions of the subject (151). And he ends the investigations explicitly stating that through experimental method, the approaches that psychology makes to behaviour are a 'conceptual confusion' that results in 'barrenness'. This statement derives from the lengthy Section ii.xi which discusses the difference between seeing and 'seeing as', and which also provides Janik with a basis for distinguishing constitutive from regulative behaviour in terms of analogical from digital structure. Evaluation and judgement arise not from explicit rules but from the imitation of a practice by which we learn what to see.[76] Tilghman uses this distinction to pursue questions of articulation and to define 'seeing as' as a practical knowing that is 'usually the ability to use the appropriate language and/or make appropriate comparisons'.[77] What Wittgenstein leaves us with is an uneasy indication that the socially 'proper' and 'appropriate' slide under pressures of time, use and history, into nominalisations. As he notes in *The Blue and Brown Books* (1952), analogies can outstay their appropriateness and become misleading.[78]

POLANYI

The focus on Wittgenstein and away from Michael Polanyi, who is often considered as the writer who opened up a wider discussion of the topic of tacit knowledge, is ostensibly because Polanyi is considered 'vague' or even 'mystical'.[79] But this apprehension derives in part from Polanyi's stress on modern science rather than the arts and humanities, which means that he is not immediately concerned with the appropriate representation for communicating to a wide public, but with the appropriate experiment which allows the individual to articulate physical reality. In other words he is concerned with a different referent for the unarticulated tacit knowledge, and with a different relationship to the audience. Since AI is a field of knowledge concerned with representation, Wittgenstein is highly relevant. However, AI gains much of its impetus directly from scientific discourse, and often takes that discourse as a first-order representation of the knowledge of modern science, with the result that it bases much of its articulation on regulative rules. Hence Polanyi's careful study of the difficulties of finding a representation for scientific experiment is also highly relevant.

In *The Tacit Dimension* Polanyi talks about scientific discovery as being guided by the sense of a presence of a hidden reality whose discovery 'terminates and satisfies'. In doing so, 'it claims to make contact with reality'.[80] What is interesting is that Polanyi's sense is congruent with Wittgenstein's on philosophy: that discovery is a series of hunches that are recognised as problems in the 'performance' or physical practice of the experiment, by the experimenter/scientist. This sense of the scientist being reflexive by looking for problems rather than solutions, and of science as performance rather than objective inquiry, forms the basis for Lyotard's discussion of science twenty years later, as the central activity of the postmodern condition. But Polanyi carefully notes the long history of this activity. He also outlines the problems that this reflexiveness inevitably meets because it lacks connection with the social that a concern with the textuality of representation ensures.

DOUBLETHINK OF POSTMODERN PLURALISM: ARBITRARY AND ABSOLUTE, OR PRIVATE AND SYSTEMATIC

Lack of effective social mediation leaves the scientist grasping after private truths, often called beauty,[81] or the enormous satisfaction and pleasure of the systematic coherence of ideology. The isolated scientist can never know whether the discovery of experimental practice is subjective truth about actuality, or has in effect suddenly made sense of a pre-existing system of epistemology. In other words, the scientist can never answer the question of whether the inarticulable tacit knowledge has just been (partially) articulated or whether the experiment has just got around to representing something within the discourse domain that no-one else has yet bothered with. Any subjective pursuit of truth will end with this dilemma about the arbitrary or the absolute. Postmodernism engages in the doublethink so well described by George Orwell: of the arbitrary as absolute, or the absolute as arbitrary.

There have been attempts to rescue the 'discovery' of science by relocating it firmly away from the cognitivist 'algorithms' of computing, into the 'beauty' and 'mystery' of mathematics. R. Penrose attempts this restatement of classic private truth and beauty, in order to point out the delusions of the 'strong AI' claims that 'the information content' of individuals 'can be translated...intact',[82] that AI is adequate 'for the description of brains and minds' (402). Penrose underlines the view that computing is algorithmic, following rules mindlessly, and incapable of evaluation, judgement of truth, or artistic appraisal; and this is an approach for which he has been taken to task.[83] Here the emphasis is rather on the private and egocentric grounds for the representation involved in his definition of creative mathematics. He claims that the 'evaluation of an algorithm can evoke *conscious awareness*' (447), because the consciousness of mathematical truth is a matter of instantaneous 'seeing' that involves 'inspiration, insight, and originality': the 'strong conviction of the *validity* of a flash of inspiration...is very closely bound up with its aesthetic qualities' (421). It is suggested that a mathematical truth is not information but a discovery akin to Plato's 'remembering' – that the mathematical concept 'is, in a sense *already present* in the mind' (429). It is an absolute. In a parallel use of Plato and mathematics that is directed toward tacit knowledge but in the name of an emergent use of chaos theory, M. Serres notes:

> If mathematics arose one day from certain techniques it was surely by making explicit this implicit knowledge. That there is a theme of secrecy in the artisans' tradition probably signifies that this secret is a secret for everybody, including the master.[84]

The reference to Plato and memory is significant because remembering what is 'already present' is with Penrose narcissistically still limited to the isolated individual, and does not take on the rather more active sense of re-membering, re-building the body, that occurs when the individual comes into contact with society, that is also offered by Plato. The narcissistic image is the other face of the psychoanalytical vocabulary that recasts beauty as the nostalgic recognition of the self as subject in the symbolic system of ideology where truth is a designed, interconnected network, and is paired with the cynical recognition of the difference of the self from the symbolic system.[85] Significantly, both narcissism and cynicism depend upon the private individual being empowered with respect to the institutional structures of the public world. Postmodern plurality is never challenged by itself, only by the different communities that it excludes and which may not be tolerant of its naive arrogance.[86]

Penrose tries to distinguish the individual discovery of the 'already present', fixed truth, from the sense of the brain 'observing itself' whenever a new perception emerges that is offered by systematic coherence. However, the two are the alternating sides of the relativist and the determinist, the private revolution and the totalising system, neither of which acknowledge the difficulty of dealing with the real, which cannot be approached without social interaction. Furthermore, once either approach begins to question its own methodology, it cannot cease because there is no sense of mate-rial/social need that can appropriately interrupt and focus on action.[87] What often happens instead is that the demands of commodification and capitalism intervene, turning the scientist or science into fame or technology; this was referred to in Chapter 1 as the anchor of profit and state patronage. Both fame and technology claim to satisfy desire by closing the circle of knowledge, so they immediately turn the constitutive rules of performance into regulative games in which the rules may be completed.[88] This is the precise way that commodification offers the seductions of power to both speaker and audience. Plato equates it with paying money for love (not sex).

Polanyi attempts to differentiate between this relativist/determinist double-think and science's apprehended contact with the 'real', by way of the disturbance that the discovery causes, saying that discoveries are 'most real' when they manifest themselves in the future in the largest range of indeter-minate results.[89] But Polanyi also realises that the difficulty of distinguishing, in the elation of discovery, between the joy of an appropriate articulation for tacit knowledge and the 'termination and satisfaction' of power either individual or systematic, is the condition of a self-validating world of knowledge isolated from society. He adds that the scientist can utter 'no...more than a responsible commitment' (78) to the truth, and that the universal claims of science are not determinist but of 'universal *intent*' (78).

In his conclusion Polanyi notes that because the heuristic field of science is not stable but problematic, the scientific community must be 'spontaneously established by self-coordination' (92) or it will lead to a fragmented society 'adrift, irresponsible, selfish, apparently chaotic', where each scientist is responsible only to their own small specialisation. These comments neatly describe the effects of postmodernism which Rorty would like to celebrate in the name of bourgeois liberalism; which Gadamer tries to harness into self-coordination while failing to address the dilemma of distinguishing between plausible and probable grounds (between prejudices and traditions); which Lyotard leaves adrift in the melancholy fragments of cynicism and nostalgia: the mirror of Baudrillard's 'simulacrum'; and which Habermas would apparently like to address by reintroducing enlightenment ideals.

If tacit knowledge for science is knowledge of the reality of the physical world, for the arts and humanities it is knowledge about the world of human interaction. Any common grounds are shared with other human beings as the basis for all types of social action. Any attempt at new common grounds for unarticulated knowledge must involve other human beings. For the arts and humanities, the sense of suddenly sharing a new common ground with other human beings is part of the knowledge being articulated. The articulation and the making public are the same thing, although it could sensibly be argued that the social sciences stress a reality about human beings that is described in the repetition of appropriate representations, and the critical sciences/arts stress the contact with reality that is learned by repeating the performed moves in processes of imitation as if they were our own. In both stresses there has been a traditional focus on providing contexts for representations, rather than delineating methodologies, although the 'scienticist' drive of the humanities in the twentieth century has been to take on methodology enthusiastically. While these approaches have opened up the variety of ways in which we study and understand other human beings, the methodology has frequently been used as an end in itself, falling into the same trap as AI of taking the methodologies themselves as the reality to be represented. This has derailed much work from questions of reflexiveness into similar consolations of power offered by the commodities of fame and of the technology of criticism, with the result that the power has been left without appropriate context. It is this that has led the arts and humanities into ahistorical postmodernism, and generated what has come to be known as the 'legitimation crisis' in all disciplines of western study.

Gadamer, Habermas, Lyotard, and more recently Rorty, offer versions of the same crisis, stemming from the same doublethink of the private and systematic, the arbitrary and the absolute. Gadamer makes the distinction between *techne* as the knowledge to be found in application, or tacit knowledge, and 'knowledge about what may be said and taught in general about such knowing'.[90] For him, hermeneutics is the art of bringing the tacit into speech, not merely through the use of technique but through practice and politics – what he elsewhere calls solidarity, 'Practice is conducting oneself and acting in solidarity. Solidarity, however, is the decisive condition and basis of all human reason' (87). Gadamer's precision is helpful. He argues against a notion of essential subjectivity and toward dialogue, a 'communal language [that] is shaped in it beyond the explicit awareness of the individual speaker', yet which offers a 'disclosure of being' founded on practice (57). The self is never transparent or present, but always 'on-the-way' (103). In the dialogue with community that makes a self-on-the-way possible, 'We are continually shaping a common perspective when we speak a common language and so are active participants in the communality of our experience of the world' (110).

But, as helpful as Gadamer may be in addressing limits rather than universals, communities rather than fragments, there is an apparently uncritical belief in the efficacy of community against ideology, and in the human instincts that construct those communities. On the one hand, Gadamer notes that 'we are already shaped by the normative images or ideas in the light of which we have been brought up', and that these norms are to be critiqued because social life is a process of transformation (135). On the other hand, these 'dominating' norms are 'taken for granted and lie...fully beyond the explicit consciousness of anyone' (82). The essay 'What is practice? The conditions of social reason', also comments on the rhetoric of community making:

> The more what is desirable is displayed for all in a way that *is convincing to all*, the more those involved discover themselves in this common reality; and to that extent human beings possess freedom in the positive sense, they have their true identity in that common reality.
>
> (77, my emphasis)

He provides local reasons for reflexively finding and critiquing common grounds, but not for overthrowing them because implicitly he does not believe one can ever get outside them: there is no discussion of the pressure of contradiction, of the need to maintain common ground for taking decisive actions, of the problems of competing common grounds, and the problem of not wanting to or not being able to critique the common grounds that satisfy desire. As a result, beauty, which he holds to be 'excessive' to survival, becomes the elation of recognising common reality. But there is no way of differentiating this 'reality' from ideology, no way of distinguishing the probable from the plausible; nor is there any way of communicating among groups that have different norms.

Habermas operates from the other extreme, because he is concerned with the need to be able to overthrow a 'blindly accepted' common ground characterised as a 'false consensus'. He distinguishes between 'good' and 'socially successful' argument,[91] on the basis of whether the argument is questioned and defended or merely influential. However, this offer of reflexive questioning as the way to distinguishing is not enough. An 'influential' argument may be challenged and then defended, while a 'good' argument may have to change with social context. His distinction can in the end only be maintained by calling on 'universals'. Lyotard simply gets rid of the terms 'necessity' and 'universal', leaving the crisis of legitimation with the incessant reflexiveness of a game that shuttles back and forth between the isolated private world and the systematic or universal.

What is interesting about Lyotard's reductiveness is that it underlines the central issue of evaluation: how to account for, assess and act on perceived contact with or knowledge of the real. There is no room for practical or tacit knowledge in Lyotard's version of 'paralogy', plurality of worlds; and this is primarily because there is no belief in a contact with the 'real'. All representation operates within paralogial structures of private or systematic worlds. A sociological version of this account precisely replicates Lyotard's rhetoric. In a paper analysing the 'hard' representation of science by sociologists of science criticising 'hard' representations of nature, Steven Woolgar maps out three stages in sociology.[92] The first is 'instrumental', the second is 'interpretivist', and the third is 'reflexive'. The tripartite categories roughly approximate the divisions in rhetoric between technique, strategy or ethos, and stance. However, what rhetorical study emphasises is that while consciousness of stance encourages reflexivity it is not automatically something to be 'proud of' as Woolgar would have it. Reflexivity does insist on placing the 'object' (knowledge, science, a person) within the analyst's own social forces, but the concept of stance adds that this is not a unitary activity: there are different ends and values involved in socialising the context for an object. The problem is not just 'to become reflexive' but to decide at which point it is appropriate to halt reflexivity in order to take decisions and effect social policy.

Here it can be appropriate to refer to the story of Pygmalion: the continuously reflexive mind, endlessly deconstructive and sceptical, is implicitly utopian, searching without end for the ideal: that Pygmalion will be able to make Galatea, articulate the reality perfectly. Yet Pygmalion needed the gods to start and to stop his articulation. To ignore the gods of biology, or any other 'universal', does not suddenly legitimate Pygmalion's activities by conferring authority on him alone. Indeed, it merely removes the current rules of thumb about reality, and leaves Pygmalion playing arbitrary games with Galatea within a self-enclosed monomaniac world.

AI AND REPRESENTATION

One reason that theorists/philosophers such as Gadamer and Habermas hang on to instinctual necessities and universals, is that they perceive the need for social action. The whole process of articulating tacit knowledge is at the root of the liberatory ethos of modern science, as it is for the humanities. What goes on in the articulation of tacit knowledge is the making public of the ways in which, or the points at which, we have decided to halt reflexivity and take action. In any scientific approach, partly because the communication of experimental work is already at one remove from the actual, and partly because of its methodological emphasis, the moment that the knowledge derived through practice is articulated it appears to make claims on fixed truth. But what it is in effect doing is making claims on reality, and thereby intervening in it and changing it. Looking at how knowledge is made public can provide guides to finding the 'appropriate' places to halt reflexivity.

The process of making public any claims on reality does not, for scientific approaches, occur in the medium that the science works. But science still needs public discourse: hence the burden of proof and repetition to involve other people in the contact with the real. Articulation of any working knowledge only happens as a result of common practice, otherwise there would be no-one to recognise appropriate common grounds. Science, specifically, works in small communities where 'common practice' is the first mode of proof.[93] The articulation puts that practice into a representation so that it can then be learned by a wider public; but all too often, for reasons previously outlined, the articulation uses strategies that aim at 'truth'. Curiously, this mimics the particular politics of authoritarianism that arose in the twentieth century in response to the massive number of newly enfranchised people. The stance is an extension of fantasy that aims to satisfy representation by providing it with the self-validating terms of tautological structures. It frequently fails and results in force when presented to an audience which does not accept or does not understand its basic assumptions: as all those who have attempted to assemble something from technical instructions will know. But science need not communicate this way. The knowledge it is about may be learned in quite different ways depending upon the articulation and particularly upon the stance. AI could be a helpful place to develop and assess more interesting approaches to scientific articulation and the effects it has on scientific practice.

The array of dyads offered by commentators on AI and representation: constitutive/regulative, mathematical/algorithmic, tacit/predicative and systematic, analogical/digital, narrative/propositional, etc., marks each first term as typical of the moment of elation where we think we touch reality. Strictly speaking, the dyads are different in kind. While they present a split perspective on language where the second term implicitly suggests that it is an inadequate code for representing the world, the constitutive or tacit knowledge of working practice is articulated via mathematical, analogical, narrative representations, that may become algorithmic, digital or propositional, and lead to regulative and predicative/systematic rules. However, there is no *necessity* for the elision from say analogical to digital, but rather there is varying historical and social pressure. For example, if you live in a society with poorly developed practices for reading analogy into the large number of figures of which metaphor, metonymy, simile, allegory and so on, are but a few, then analogy will slip swiftly into digital analogue. Different articulations effect/affect practice in differing ways dependent on this pressure.

Once articulated, practical knowledge need not be merely relegated to rule-bound system, because the text of the articulation can permit the repetition of the performance, practice or work that insisted on the articulation in the first place. However, at the same time as it makes it possible, articulated practical knowledge does inhibit the articulation of other practices. Some ways of representing will be closed off, and others will be opened up. So it becomes necessary to be reflexive about the representation, to understand what our position is with respect to it and how we are participating in it, not so that we can engage in the incessant shuttle between the private and the systematic (whether universal, ethnocentric or ideological), but so that we can connect the articulated with the unarticulated and take a step toward evaluating appropriateness.

To answer the questions of how much and where we are participating in representation, and what any articulation makes it difficult to say, we have to address the common grounds in the practice of that representation. Nearly all the commentators interested in AI and society, whether or not they are concerned with tacit knowledge, discuss these questions of representation. The point in each case is to underline the need for constitutive practice, the first term within each dyad of representation. A contingent concern, not always followed up, is the need for reflexiveness that is the initial step to constitutive practice. But constitutive practice also involves an understanding of rhetorical stance which brings together the strategies of reflexiveness with its own critique of articulation; it brings together strategy and the material; it focuses precisely on the elision from constitutive to regulative and asks us to be clear about position, about where to stop reflexiveness and take action.

The stance of AI has predominantly been based on the self-enclosed rhetoric of scientific communication. It moves inexorably around the structures of tautology-making that scientific methodology has elaborated, taking common grounds as axiomatic, as 'agreement', as 'best-fit', as oppositional, as satisfactory, as unnerving; and in this way raises peculiarly acutely the problems inherent in fantasy. The stance of fantasy as elaborated above is a way of dealing with unarticulated knowledge and social context, by treating all common grounds as systematic yet choosing simultaneously to forget this strategy. Once the system has been defined, it can satisfy any desire articulated within its common grounds. It can offer continual repetition of that satisfaction, it becomes addictive; and the compulsive addiction of computer-worlds is well documented.

A recognition of this self-deluding procedure leads R. Brooks to try completely to evade the need for representation in AI. He concludes that

> When we examine very simple level intelligence we find that explicit representations and models of the world simply get in the way. It turns out to be better to use the world as its own model.[94]

What this reveals is that he takes 'representation' to be explicit and so of course finds it inadequate. Furthermore, using the world's 'own' models necessarily involves articulation of another representation. As he proceeds, Brooks appears to become aware of this. He becomes insistent upon 'traditional' representation being the root of the problem, and he turns elsewhere inventing in his notion of mobotics interesting new possibilities for AI. But as Kirsh's commentary indicates, these have their own implicit representations.[95] Any evasion of representation on the grounds that it is inadequate to the world is tautological, because it is based on the concept of a linguistic adequacy which fails by definition. Kirsh points out that the idea of an inadequacy of axiomatic knowledge is also bound up in an essentialism based on egocentricity: the denial of a public and social space.[96] These concerns parallel the broader ethnocentricity of Rorty, with his similar evasion of inadequate representation. And Brooks' comment that he is not interested in the significance of the implications of his work, indicates a similar evasion of social concern.

The kind of satisfying repetition enabled by chosen or evaded tautological worlds is quite different from the repetition enabled by the constitutive practice or performance, because the common grounds of performance can never be defined within a fixed system. This realisation forms the centre of W. Clancey's monograph, 'Model construction operators',[97] about which he comments,

> I realised that many confusions about representations could be resolved if we see them as alternative perspectives on a single 'virtual' formal system.... I realised that representations and reasoning processes that were commonly viewed as different could be related by a shift in visualization.[98]

Yet Clancey's commentary on the way his approaches have developed, almost exactly parallels the shift in scientific concerns with representation from maps to the plurality of postmodern design and its attendant problems, and does not look any further. Some researchers are turning to alternative ways of agreeing upon grounds for definition and upon grounds for reference.[99] Rather than move directly into the plural versions of postmodern representation which, as the commentators on representation in science have recognised, are still quite separate from society, perhaps AI could consider using the concepts of rhetorical stance present in the humanities to pursue alternative ways of agreeing upon the grounds for the representation it articulates.

Were AI to investigate the constitutive rules of topical and analogical reasoning, it might be able to offer science a mode of proof and repetition that reintroduces the phronesis of science into its representation, reintroduces the probable and insists on the textuality of public communication.[100] But if AI is to duplicate in virtual reality the repetition of satisfaction that is the representation of science (by technology) in mechanical reality, then it will place itself within the story of Frankenstein. Unlike Pygmalion's Galatea, Frankenstein's creation is man-made with naive arrogance. It takes the second-order representation of science and makes it literally real; technology is born and there are no gods to get rid of it. It is helpful to remember that Pygmalion is a myth by man for men, underlining a belief in their own god-like actions, while Frankenstein and the monster are part of a myth written by a woman looking at a man's world, and providing a devastating critique of the literalising of fantasy.

4   The socialising of context
Methodologies for hypertext

In application, the kind of knowledge the computer encourages is rational-ist, linear and analytic, mimicking the public communication of science. This was a widely acknowledged desideratum of most humanities' use of the technology during the 1970s and 1980s, and is apparent even in the textbooks of artificial intelligence, which as a discipline has a research base that continually questions the axioms or assumptions that underlie such 'formal' reasoning. Research-based AI and its philosophical commentators have moved consistently into attempts to provide for programming in particular an environment or a set of surroundings which is usually referred to as a 'context'. Similarly, although only in the past decade because it was responding to a new technological advance, humanities use has turned toward hypertext as a medium that can offer 'context' and allow one to narrate the story of a knowledge in a far more flexible way.

Artificial intelligence researchers have moved toward context in order to offer models and analogies for representing expertise rather than explicit 'ends' and 'proofs', and they overtly discuss such expertise in terms of the tacit knowledge of science. Yet as the previous chapter intended to foreground, that tacit knowledge is wrapped up in a specific understanding of representation, language and ways of knowing, that is complicit in the terms of the autonomous individual within the liberal democratic contract. The 'tacit' is primarily the unspoken set of assumptions of science for which one needs context in order to foreground. It is also, to a smaller extent, the idea of skill or expertise that is impossible to represent because language is inadequate and therefore necessarily tacit. Within this understanding, all language and all representation inevitably points to an absent fullness or presence. Existing alongside these implicitly transcendent notions is the idealism of 'beauty' as a wresting of the actual into representational fullness or truth: an action that is carried out only by great scientists, who become great through their ability to produce these true representations.

Computing science repeats this structure of complicity in liberal demo-cratic knowledge in a fundamentally different way, which Sherry Turkle refers to as the 'intellectual hubris' of taking simulation to be real.[1] Computer programming allows for the construction of fantasy worlds within which one is tested, often violently, as one moves toward the goal in order to 'win'. Yet all too often this is enabled because of the abstract and formal structure of the logic that is mechanistic and linear, allowing one to disregard the primary assumptions of the game and to gain a sense of mastery as one moves from stage to stage. In *The Second Self*[2] Turkle makes the now widely recognised distinction between 'hard' and 'soft' programming. While the formal is hard, soft programming is based on object relations; it is interactive and conversational, concrete, and close to Lévi-Strauss' concept of the artist as bricoleur. Slightly later (1990), in an article written jointly by Turkle and S. Papert, this distinction has changed its structure.[3] Formal logic is no longer a stage but a style of reasoning (143), and computing is seen, as in much AI research, as working right along the borderline between the abstract idea and the concrete physical object (131).

The later article elaborates on the idea of programming styles leading to epistemological pluralism. The formal style takes objects as properties (143), is suited to distant planning, and results in an analytic rule-bound strategy (148) that is fundamentally algebraic (155). The formal, just as Carol Pateman describes in the 'objectivity' of the infinitely replicable autonomous being which attests to the loss of the individual self, creates a 'decentred' self, the human individual missing from this kind of knowledge. In contrast, Turkle and Papert offer the bricoleur who takes objects as relations (143), is suited to working closely with the issues, through contextual and associational strategies (148) that emphasise biological and social factors. Yet while bricolage claims agency, it claims it here simply as another 'style' without pursuing the historically specific issues of domestic repression and coercion to civic consent that have made it both inappropriate in terms of public politics and open to its own particular extents and limitations.

The detailing of bricolage as an alternative computing strategy is similar to the gesture that philosophers and teachers of science and computing make to the arts and humanities. Turkle and Papert indicate how computing, lying as it does on the borderline between the abstract and the concrete, is dealing largely in issues of representation. They make the point that science is concrete but 'canonically' must be represented as formal (130). Yet one of their immediate examples is how computing can move from the verbal and abstract idea to the visual and concrete: from the word to the icon: from the absolute to the analogy. The problem here is that all media have rhetorical techniques and strategies that are equally problematic or enabling, depending upon specific historical and social materialities. The icon can be just as absolute, if not more so, than the alphabetic word, and for a variety of reasons. Computing is indeed a medium and has textuality, but no one technique or strategy will necessarily be appropriate to a particular kind of knowledge. As indicated at points throughout the three preceding chapters, linear, rational, analytic logic in the sixteenth and seventeenth centuries had democratising as well as spiritual and commercial initiatives encouraging it. It could speak across educational, social and cultural elites and replaced

Latin as the language of legitimacy – only to form its own elite due to different historical conditions to which no doubt it contributed.

What this chapter will attempt to do is look at the ways in which the humanities' turn toward hypertext often claims for that medium the same privileges that scientists make when they gesture toward the arts, and which similarly only go part of the way toward understanding textuality. In particular, I will look at the claims that hypertext can mimic oral strategies and is therefore more interactive and personal, and that because it provides context it resolves the restrictions of formal programming. This concept duplicates a prevalent understanding in the arts and humanities that the oral is more flexible and engaging, and less authoritative, than the written. Much contemporary debate over the issue once more implicitly assumes that language is (in)adequate: hence the codes of second-order textuality and the idea that the only first-order textuality is transgressive or transcendent.

The study of hypertext structures will underline an understanding of techniques and strategies as never enclosing or isolating in themselves, and open not only to transcendence and transgression but also to engaged interaction that works with the textuality in specific situations. To do so I will look first at the background for these issues in the history of rhetoric, and then at the relevance of contemporary debates about legitimacy, representation and epistemology. Although the focus will be on hypertext because this is my own area of practical work, I would suggest that the discussion of the extents and limitations of its textuality could be extended to other models of soft programming. For clarity's sake, while I intend to criticise the claims made for hypertext, I would like here at the start to say that the critique aims to open out the possibilities of the medium, not to close them down.

TOPICAL REASONING IN THE HISTORY OF RHETORIC

Topical reasoning in social persuasion: consensus, corporate and authoritarian

Western European ways of organising information, creating data and ordering knowledge from the classical period until the Renaissance, may be read as based on the categories and structures of Aristotle's *Topica*. Both the categories and the structures are familiar to us today as contexts of identity and of structural relativity. The categories are: essence, quantity, quality, relation, place, time, position, state, active/passive; and the structures comprise: commonplaces or grounds for prediction of accident or happening, rules for the comprehensive evaluation of two or more predicates, predictions of genus and property, and definition. As noted earlier, Aristotle allied the topics with the validation of the non-social argument of philosophy and science, as well as with the social practices of rhetoric which included the strategies of the rhetor and the audience, or ethos and pathos, as topics in themselves. But whereas I previously looked at some of the implications of opening the door on to non-contextual reasoning, what I would like to emphasise here is that there were on-going implications for social persuasion.

Within a rhetorical analysis of any occasion, ethos works like a topos: a common ground on which, or from which, argument and discussion can proceed. As social action in *Politics* and as persuasive strategy in *Rhetoric*, Aristotle describes the shift in his society from aristocracy to constitutionally agreed state; or, as he also puts it negatively, from oligarchy to democracy without a constitution. The shift takes place against a concept of government that is tripartite: government by one, by the few and by many.[4] Those who govern need to persuade the people over whom they rule and especially whom they represent, of their legitimacy, and Aristotle elaborates on the different kinds of ethos that a rhetor may take up within any one of the three types of government.

For example, government by one person may be a monarchy dependent upon the complete authority of the king, or it may be a tyranny in which authority is derived from forcible coercion by the one over all others. Government by the few may be aristocratic, with an ethos dependent upon legitimation through educating all others in a particular way that underwrites and supports the aims of the aristocratic system; the ethos is dependent on strategies described in common interpretations of the *Topics* as evading contradiction, in order to construct a stable system. Yet government by the few may also be oligarchical and focus on creating a stable system at all costs in order to gain wealth for those in power. Just so, government by the many may be carried out constitutionally, so that all people have redress to certain rights that ensure equability between communal and individual good. In this case the ethos of the government allows for the many voices of democracy to which Aristotle accords the value of difference and contestation – the example he gives is the understanding of music,[5] which is significantly important to a political education[6] – yet constitutional democracy also provides a set of agreed rights that constitute guidelines for the resolution of contradiction in order to take political action. Without constitutional guidelines democracy would devolve into populism, or what is now called ethnocentricity, which is closely related to the Orwellian slogan that 'All people are equal, but some are more equal than others'.[7]

In each case, if the government aims toward private interest then it begins to participate in inappropriate or unethical rhetorics. The common factor of each politically abusive rhetoric is its attempt to work within a set of enclosed grounds or topics, as if they were already agreed to *a priori*. The rules of the tyrant, the oligarchy and the democracy are based on a prior assumption of the good of individual gain, and their ethos is not to discuss but to succeed in gaining assent to their individual agenda by force, or by depriving the governed and the represented of the information or the education that will allow them to contest that agenda. Aristotle notes that to get the populace to legitimate rule under a government aiming at individual gain, is increasingly difficult as one moves from democracy where the legitimation is more accepted because many people seek individual gain, to oligarchy and to tyranny, with the result that the latter two are very short-lived.[8]

On the other hand, in Aristotle's writing the rule of monarch, aristocracy and constitutional democracy have embedded within their rhetoric a relationship with the governed or represented which invites active assent about what constitutes 'good' for the society as a whole. This rhetoric necessitates a different kind of ethos: one that is to various extents open to discussion, although there is a larger of range of ethos positions for a democracy than an aristocracy than a monarchy. The monarch, for example, cannot be 'systematic' since legitimation is not dependent on others agreeing, nor can a monarch or aristocracy use broad consensus since there is no other person/people of equal status with whom to agree. Here the *Topics* are open to a rather different interpretation than as a set of prescriptive rules, for given Aristotle's statement that rhetorical debate is necessary for considered social discussion, the *Topics* can be read as guidelines for the resolution of debate into agreement on grounds, which agreement is essential for decision and political action.[9] This shift is in effect a result of moving a topic from the realm of science, where it functions to introduce and reinforce *a priori*, to the realm of the social and political where it functions to persuade and legitimate.

What is being described is a consensual ethos moving toward corporate decision-making. I take consensual rhetoric to signify debate and contradiction until agreement can be reached on action, not, as some people would have it, that consensus is habitual or conventional agreement.[10] Consensual agreements, however, can with repetition elide into the conventional and that other sense of 'corporate': not as embodying and holistic but as systematic and static.[11] Indeed, democracy, oligarchy and tyranny can each function effectively within this second sense of corporate ethos because it discourages debate and disagreement. However, if the ethos is de-legitimated for some reason, these forms of government usually have few persuasive fall-back positions because they do not encourage active modes of assent, and frequently have to resort to force: Aristotle distinguishes the demagogue from the orator just at this point, saying that the orator does not know enough about force.[12]

As this interpretation indicates, I do not read Aristotle only as laying down prescriptive rules. Sometimes he does, for example when he simply states the inferiority of women as an *a priori* in *Politics*, and then moves on to argue a catalogue of calumnies. However, his writing engages dialectically precisely with those issues over which he perceives disagreement about *a priori* within his intellectual community.[13] Aristotle's philosopher describes the rules of thumb for social action within the restricted slave state that Athens was at the time; he also describes the rules of action for the stable sciences, and the actions of authority. He does so demonstratively not rhetorically, because he describes a metadiscourse that has been separated from social action.[14] But, the moment the 'rules' enter social immediacy, the moment the philosopher becomes a rhetor, the topics are immediately recast into the materiality of politics. Aristotle notes at the end of the *Topics* that the philosopher need only know the categories and essences because he is investigating 'by himself'; but the person who wishes to enter social life needs the skills of the rhetor because 'reference to another party is involved'.[15]

In my reading, Plato is less willing to separate the philosopher from social action, theory from practice. Plato distinguishes between the philosopher as articulating the point of radical disjunction, and the sciences and the construction of social negotiations around them.[16] He distinguishes in *Phaedrus* among rhetoric as merely technical display, and rhetoric as coopted and compromised by systematic convention, which he symbolises through the metaphor of exchange,[17] and rhetoric as philosophic, to which he assigns the metaphors of writing, medicine and gardening, and by analogy the field of morally sound oratory. If one adds the description from *Gorgias* of the ethos of force, or 'might is right', which Socrates does not take to be an ethos at all but a non-rhetorical coercion, Plato leaves us with the description of four ethos positions: the forceful, the narcissistic which is dependent on technical display, the systematic or corporate dependent on strategy, and the philosophic which is engaged in consensual negotiation with a specific set of conditions. It is of interest that he discusses the first three at length in his many works but rarely the last, although he writes engagement into his rhetoric on many occasions.

It will be apparent from these readings that I find much in Plato and Aristotle that others in recent criticism would attribute to the sophists.[18] There is the possibility that the sophists needed to be reclaimed in order to rectify a modern preoccupation with Aristotle as prescriptive and Plato as an essentialist idealist, with neither of which interpretations would I agree. The reclamation of the variety of sophistic techniques and strategies underwrites the sense offered here of Aristotle's flexibility in his rules of thumb for persuasion, and the sense of Plato's moral particularity as he proffers the 'ideal' in order to insist on the necessary limitations of articulating the real and ways of negotiating and re-negotiating those limitations. Indeed, C. J. Swearingen's illuminating *Rhetoric and Irony* extends and elaborates upon this sense of an 'antistrophe...[to] the monological'[19] that the texts of these two monoliths of western philosophy ironically convey. The sophists appear to have contributed to the political discourse of fifth- to fourth-century Athens, a series of techniques and strategies that allowed and possibly instigated the suddenly plural voice of the restricted democracy that emerged. These varied people had very different interests in power, and given access to power needed to establish ways of contesting their claims. What Plato and Aristotle seem to have developed[20] is the realisation that continued competition and contradiction need at least temporary arrest to enable action; they then proceed to focus on how to achieve this arrest appropriately, or for the common good, 'common' of course to citizens of a slave state with only partial enfranchisement.[21]

If Plato denies the value of all non-social topical reasoning while Aristotle claims that it has value in certain circumstances, the differences that emerge between these discussions of argument and reasoning enact the history of the topics. They are at times to do solely with category and/or structure as they are in mathematics today,[22] where their significance is often neither probable nor analogous but fixed within a determined system: for example, the analogue-makers of artificial intelligence restrict themselves largely to the structures of predication in the topics. At other times the topics have been reduced simply to questions of person and act, as they were in schools of eloquence or as they often are in the construction of political personality. This reduction to ethos or pathos or res, and denial or argument, leaves no place for questions of reasonable validity. But the topics have also yielded a productive combination of the two discussions in Ciceronian rhetoric,[23] versions of humanism,[24] the new argumentation,[25] and in recent critical discussions within feminism, which I will further discuss in the following chapter, some combinations of which display a rare and profound belief in the impossibility of separating the social from the topical or contextual, and a concomitant difficulty in finding a vocabulary for speaking about the resultant holistic materiality.

Topical reasoning in the oral and the written: correct/corrupt or adequate/inadequate language

The classical focus on topos or the topoi within social persuasion was concerned with ways of structuring arguments as analogical reasoning from a probable rather than from an *a priori* basis. A topos provides a general setting for a discussion, a framework for arguments rather than a fixed set of rules, standards or axioms. Those involved in the discussion need to agree that the setting is appropriate, and the argument proceeds within social and historical materiality. Traditionally topical or contextual reasoning appeals to belief, sometimes to passion and emotion, but never to intellectual gameplaying by proof. It is dependent for its full range on a combination of the categories and structures with the topics of person and act. Discussion from common grounds is based on persuasion to the probable and the reaching of consensus. The problem with consensus is that while its terms are initially agreed to by 'consenting adults' who are aware of the rules of the linguistic and social context, that consensus is all too often allowed to maintain its 'agreement' beyond the relevant context. Despite the impetus from a social or historical event that generates the initial consensus agreement about the common ground or appropriate context for a topic, those topics tend to accrete into corporately held agreements. The engaged consensus rarely maintains its sense of person and act for long, partly because the corporate is more stable and can maintain the status quo needed by the political structure. The heated debates between rhetoric and logic that occurred throughout the two thousand years following the *Topica* acquired a distinctive turn with the Renaissance and the increasing use that print gave to the written medium, and resulted in a radical revision of the topics, particularly those of person and act.

Renaissance attempts to resolve the problem, in other words the techniques, strategies and genres that were devised to deal with social and historical materiality, were bound into the changing world of the printed book: it is the increased audience, the shift in cognitive skills that print asks for from a manuscript- or oral-based culture,[26] and the practicalities of communication in the print medium which radically redefine the community of scribe and script into that of writer, printer, reader and the book trade.[27] Many studies of hypertext, just as do many commentaries on scientific communication, return to the Renaissance, specifically the seventeenth century in England, to ask questions about the implications of the representative media of the oral and the written. Yet without an understanding of the history of rhetoric, it is often assumed that the predominant use of topical reasoning that provides context in oral media, automatically engages with the social. At the same time the written media, whether manuscript or print, are taken to be conducive to isolated authoritarianism because the writers are not face-to-face with their audience. Just so, many studies return to the Renaissance in their attempt to sort out the relationship between formal reasoning and the written text, and without the broader history of rhetoric there are similar misunderstandings about the way that oral and written texts interact with logic and represent knowledge. Many people have suggested that the stabilising of the topics into sets of static proofs was a result of the new medium that print offered to writing. But concern with static topics, or the *status causae*, had arisen on several occasions well before printing. Each occasion found different reasons for the loss of social materiality for the contexts being employed, and each offered different solutions. Indeed, it is likely that Erasmus found in the effects of the print medium a way of counteracting the stasis of scholastic nominalism.[28]

Taking a cue from discussions in the world of critical theory, writers on hypertext have been looking at the work of Jacques Derrida and the interpretations derived from his study of the oral and the written. Derrida's texts themselves initially draw from Plato, particularly from Plato's second work on rhetoric mentioned above, the *Phaedrus*. This work is usually treated as proof of the condemnation of writing[29] – a fact that delights those who wish to privilege speech and angers those who see in it the seeds of a western philosophical distrust of the written – but it is primarily a critique of speech. An overarching direction of the text is that rhetoric is not merely a set of techniques; it is also instruction in the strategy of speaker and audience, of ethos and pathos, and a guide to stance or 'goodness'.[30] Each element, technique, strategy and stance, is addressed distinctly; it is made clear that the discussion of technique in particular, and strategy to a lesser extent, is not focused on a debate between the oral and written as such. Rhetoric in Plato's society was not a description of some simple orality: in the *Republic* the poets are banished and the poets of Plato's time are not writers but real speakers.[31] Rhetoric was a description of oratory.

As a description of oratory, rhetoric may well be considered a description of a form of group communication, hence the importance to it of social immediacy. The writer of *Phaedrus* recognises this group positioning, and spends most of the text criticising the relationship of two orations to their audiences. But the criticism is enacted in writing, and while the techniques and strategies of oratorical rhetoric may not be suited to the written, what is said about stance lies at the foundation of both media and provides legitimation for the text Plato writes. Of course Plato did not criticise either the two orations or writing simply because they described communication to groups; what he was drawing out were the precise details whereby both orator and writer might produce texts that would debilitate their audience. In other words, issues might be put forward that both formally and structurally tried to disguise or hide their constructed status, thereby 'naturalising' them, making them seem 'the case'. It is this that is at the centre of the present debates about the recognition, and means of recognition, of ideology.

Writing primarily draws forth criticisms of strategy resulting from its absent audience. The most immediate of these criticisms is that the audience cannot ask questions; the texts cannot reply. 'If you ask them what they mean by anything they simply return the same answer over and over again'.[32] Further, because the audience is at a distance, the arguments of writing cannot be shaped with specific readers in mind: 'a writing cannot distinguish between suitable and unsuitable readers' (97). The implications of these criticisms have been taken up ever since, not just for writing, but for all systems of communication, particularly communication at a distance from its audience. In terms of technique, writing is criticised because it is a medium that is thought to destroy the need for and the exercise of memory. Readers will become forgetful; 'they will rely on writing to bring things to their remembrance by external signs instead of their own internal resources' (97) – again a fear that has accompanied the inception of all technologies ever since because it is related to the fear that wisdom will be superceded by a patina of information. At root, writing is criticised both in techniques and strategy because it deprives its audience of a basis for reasoning and evaluation. It should be underlined that what Plato is specifically attacking is the writing of speeches. At the time of *Phaedrus* this was a new if increasing practice for, generally, writing was not common. The other main area outside commerce where writing was growing and which Plato addresses implicitly, if not explicitly, was philosophy.

The most detailed criticism in the *Phaedrus* is reserved for oral speeches, although this is complicated by the fact that of the two orations studied the first is retold from a written version, and the second (although supposedly direct) reaches us through the written medium of the book. Criticism of the first speech from Lysias is made quite explicit by Socrates, who quickly uncovers the shambles of figuration and illogicality of which it is made up. The course of Socrates' own first speech is technically far more sophisti- cated, but concomitantly has its own more serious drawbacks.[33] The second speech ostensibly criticises the techniques of the first, insisting on precision in figuration, and clear analysis and synthesis within the argument so that the audience need not be told what to think but instead be led to under- stand. These aspects are the specific contribution of the art of rhetoric to vocal communication, since ambivalence and vagueness are the negative qualities of oral poetic expression.[34] But having given the second speech, Socrates stops and criticises it himself, saying that in his splendid argument portraying a love that manipulates he has purchased 'honour with men at the price of offending the gods' (44). Despite the sophistication of his technique and the shift of his strategy away from force to manipulative argument, he shows that his stance is identical to that of Lysias. The figures may be clear and the argument logical, but they delineate their own grounds and do not allow the audience to go beyond them and introduce questions that might challenge, disrupt or seriously question. The speeches deny a participation and assume the authority of the speaker over the audience.

It is a rather obvious joke that the centrepiece of the *Phaedrus* is Socrates' second speech, a written mythological poem that is like neither oratory nor writing as criticised elsewhere in the book. What the writer of *Phaedrus* is doing is presenting the reader with a work constructed in a responsible written rhetoric. The shift in stance underlines the different strategy taken and emerges in the emphasis on a rather different set of technical devices, heavily analogical, that engage the reader in dialectical involvement. In the case both of writing and of oratory, it is the effective fixity of authority which each achieves if the rhetoric is used unwisely that Plato criticises. In other words, he is criticising the authoritative stance that is the counterpart to depriving the audience of a basis for evaluative reasoning. For Plato such a basis resides in textual dialectic, which generates participation among reader, writer and writing, or among audience, rhetor and speech.

It is not the absent audience that is a problem, but the deprivation of evaluative reasoning that can apply equally to the individual or the group, to the spoken or the written, and which indicates the stance of the rhetoric. Through the analogy of the lovers which runs throughout the writing, Plato indicates that an understanding of stance is the province of the philosophi- cal lover who also comprehends strategy and technique; the nonphilosophical lover neglects stance, while the non-lover of Lysias' speech is merely a technician. This view of rhetoric as variously philosophical, sophistical and technical, and addressing itself to materiality, ethos/pathos and device respectively, is most important for an understanding of the scope of rhetoric. It makes it possible to understand that techniques may vary just as those of Socrates' first speech are of oratory and of his second are to do with writing, yet in terms of ethos, the strategy of the oral simply provides an analogy for the strategy of the written: for example, the authoritative ethos of the written speeches is no different from the ethos of writing portrayed in the story of Theuth. Stance itself is the root of the *Phaedrus*, but it remains an effect of reading the analogies; it is not conceptualised.

The *Phaedrus* is central to the language debate over the difference between the oral and the written, because the written is often cast in terms of being a second-order sign system that necessarily corrupts the spoken word. Combined with influential interpretations of Derrida's *Of Grammatology*,[35] it is the supposed attitude of both language and literature critics toward the written as 'corrupt' that has infused many arguments over the last twenty years. In that interpretative work, Derrida is understood to hold Plato, Rousseau and Saussure as examples of a quest for purity, for the natural, for presence within spoken language versus the corruption of the written. Yet Derrida himself is aware that the early twentieth-century attempt to divorce the spoken from the written was partly to do with the limitations of written grammar, the mathematical theories underlying communication, and argues the same for Plato and for Rousseau. The basis of the argument in *Of Grammatology* is that a more reasonable attitude to writing would end by bringing both the oral and the written within the broader field of grammatology or semiotics. But since the burden of his argument is to expand upon the wider implications of writing as *différance*, he goes on to emphasise that the linguists have created their own rule-bound limitations and does not follow up on what they might gain from exploring their own *différance*.

The reaction of other criticism to this eloquent work has been a huge concentration on the differences between the oral and written, rather than on their similar foundation in *différance*. Literary criticism has made much of the strong line of oral=pure and written=corrupt by which Derrida defined language criticism, rather than writing or orality in themselves. This distinction is directly part of the post-Cartesian development of all communicative language as inadequate to full representation. A language that is continually failing to achieve the purity and truth of fullness is wide open to charges of corruption and manipulation. Criticism has often converted the distinction to the argument that the written text is imposingly authoritative while the oral is dialectically engaging. This in turn has fuelled the simplistic attitude that writing is the imposition of power and that the printed word is the root of hierarchical authority, while the oral has to do with democratic power and that the spoken word is more natural. However, historians of language have told many different stories about the relative corruption or purity of both the oral and written: for example, of how the oral loses its public 'purity' and becomes merely popular just at the point that authority is being vested in written products,[36] or how the reformist Protestants construct their arguments for the purity of the written in order to legitimate not only the authority of the Bible as the word of God but the possibility of its interpretation by any individual.[37]

Assessments of the implications of print and the corruption or correctness, the inadequacy or adequacy, of writing have also engaged many commentators on issues of literacy. Here, particularly in cognitive psychology which has had a large impact on AI, the emphasis is on speech as interpersonal and writing as detached, yet also upon the need to describe writing as other than a second-order sign system. Again, the neglect of rhetoric has held back understanding of both processes, because the focus on the absent audience leads virtually all researchers to consider the differences between oral and written as located in the presence or absence of context. Because speaking is face-to-face, it is necessarily surrounded by the context of the communicative act. In contrast, writing is supposed to be context-free. This is the source for its apparent objectivity, which some take to be an effect of cognition itself. There have been many elaborations of this approach, looking at narrative, characterisation, sentence structure, cohesion and other elements,[38] and they have been particularly successful in studies of scientific and technical writing as well as journalism.

However, the qualities of 'objectivity', rational analysis and detachment described by cognitive psychologists as the result of the written in science, technology and journalism, transpose easily into the techniques of denotative language, verisimilitude, sequential linear narration, and unseen author, that are called upon by many other writers of narrative in their attempt to provide observation with context. While some writers[39] indicate that the single reading and completely controlling author are desirable, the rhetoric of argumentation points out that it is unethical. The techniques and strategies are of enormous pragmatic use in achieving short-term ends, or what cognitive psychology calls the transfer of factual information. But unless the process of that achievement is recognised for the corporate movement it is, within which individual decision and choice is temporarily deferred to another authority, then the pragmatic short term can easily elide into the dictatorial long term. In other words, although it is helpful to move on to the idea that the written does have a context, to then constrain this to the internal invented context of a possible world is not going far enough. Writing is just as socially grounded as speaking, but in a different manner. The audience is never absent. A growing amount of cognitive study is becoming more interested in this, and hence in the learned rather than inherent aspects of writing. Several studies suggest that it may be education itself that encourages the skills in logic previously attributed to writing,[40] a suggestion also elaborated and claimed as the case for theoretical logical skills.[41]

Here the history of the development of education with that of rhetoric is revealing. Western European and North American education is based on the influence of Quintilian during the Renaissance,[42] with strong Ramist overtones that separate logic from rhetoric, idea from utterance, and absolute from context.[43] Although frequently produced for ten-year-old schoolboys, sixteenth- and seventeenth-century school textbooks in writing were often little more than handbooks of devices with examples. They existed hand-in-hand with the new science and its devotion to geometry, mathematics and theoretical syllogism, put to the use of rational, analytical logic. This division curiously allied the teaching of mathematics and logic, one of the centres of the new science, with ideal, context-free writing; it also allied the teaching of eloquent rhetoric with utterance and performance. There are many other reasons for the division, but in terms of education, the nineteenth century exacerbated the division with the rhetorics of Whately (logic) and Blair (eloquence) underwriting the educational tradition that writing was either correct or corrupt, adequate or inadequate. Teaching gradually eradicated corrupt eloquence, leaving the equation literacy=writing=analytical logic, allowing the social historians and cognitive psychologists of today to think of formal schooled literacy as parallel with, it not identical to, the theoretical syllogisms of modern scientific procedure.

Part of the problem here is that contrasting this kind of literacy with conversational orality does not yield helpful comparisons. If oral skills were ever taught to a degree proportional with those of writing, specifically the kind of writing described above, useful differences might emerge. But the fact is that we educate people formally in the skills of oratory rarely if at all, whereas written skills are undertaken with intensity for a good ten years at least. When oral skills were taught with this degree of intensity against the background of scholastic rhetoric, they produced medieval dialectics that were founded on the elaboration of the theoretical syllogism. If we want to divorce analytical and categorising skills even further from writing, we could look at recent work into the classifying systems of nonliterate cultures (Berlin 1977). However, while research into the history of literacy education has done much to argue against the essentialist notion that rational analysis is inherent to writing and provides it correct form, when looked at more closely that research may be seen to reverse the axis of correct/corrupt, claiming that analytical writing somehow corrupts the genuine products of popular communication, which are in turn implicitly valorised.

Topical reasoning: querying the Renaissance

To look at the history of oral and written language, particularly the education offered in either mode, is to find that neither is inherently more correct or corrupt, more adequate or inadequate, than the other. In any specific period, people are working with social conditions and political pressures that encourage or restrict an engagement with the limitations of language. One of the main reasons that so many people interested in hypertext look as far back as the Renaissance, is that the introduction of print involved profound renegotiations with those limitations. The result was centuries of experiment with the physical and linguistic extents of printed literary products, which centred largely on how to engage the print medium audience with the social immediacies within which the topics, or now the themes, were appropriate.

The audience for printed books in the Renaissance is not the small community whose primary mode was scholastic disputation, but an enlargening group, from differing social and educational backgrounds. The Renaissance response to the ethical requirements of this multivarious virtual audience seems partly to have been a development of theory for written genres which would place and position both writer and reader; in other words, develop new topoi for person and act in written genres. For example, there is the extensive concern with what may be considered one of the most sophisticated structural attributes of the novel: the narrator. But genres in themselves do not ensure consensus, indeed they often capitalise on the success of certain patterns by repeating the technical and strategic formula of kinds of writing which have already been accepted and incorporated.

There were also special cognitive problems for the topics with the medium of print and its effect on the structure of writing; most obviously and importantly these problems raised questions for consensus and corporate reasoning in terms of repetition and cohesion. Repetition in oral and manuscript work was a device for conveying authority, but primarily allowing for variation of theme and interpretation.[44] Writing itself is not averse to repetition, but the uniformity and relative stability of the print medium allows for the development of a rational analytical model for writing, where the whole point is that you do not repeat or go back to persuade again to common grounds, but simply check back through preceding papers for the proofs for earlier premises. When repetition does occur it has to be exact because it can always be checked, and it attains validity only if it 'matches' what came before.

A sense for the stylistic implications of this technological change can be gained from a brief look at the development of indexes. The topical commentaries in the margins of medieval manuscripts gave rise in the growing authority of the text to intercalated headings often by topic.[45] When the printed book appears, with dependable repeated layout, so too do indexes which cite the topics in a list at the start or end of book; and eventually with the introduction of page numbering, indexes cite topic against page number. The topical commentary has turned into fixed category, yet the need for more relevant and immediate commentary resurfaces into marginalian comments and the keeping of commonplace books. Repetition within writing has to find other strategies than, say, lists of items for variation. If it does not, then the self-evident status of its argument becomes too clear, as when commonplace becomes cliché, character becomes melodramatic type, parable or saying becomes platitude, allegorical enigma becomes emblem, and narrative type becomes narrative stereotype. Repetition in the printed medium raises the whole question of cohesion by way of the corporate or consensual approach and response to self-evident status – whether one views it as a comforting, reinforcing ideological constraint, as merely social commonplace, or as potentially distorting manipulative requirement. Simultaneously, it also raises or foregrounds the strategies which operate by subverting the concept of 'matching' repetition, such as parody, paradox, irony and so on. Furthermore repetition, and the hundreds of rhetorical devices conventionally gathered together to describe it, not only maintain ideological representation and subvert it, but allow for work on articulation of social needs outside of the ideological in the civic and domestic.

Topical reasoning by way of commonplace books could be highly inter-active and engaged. Montaigne's method of cross-indexing ensures that any one topic is in a dynamic relation with other topics at all times. Many writers believed that the physical shape of the type on the page had persuasive effects and they worked carefully and precisely with design elements and illustration. And from the late seventeenth century the interaction of the multimedia, multiauthored texts constructed in periodical publication became and have remained the most popular reading medium for printed material. However, while these efforts were made and are on-going, from the seventeenth century they were no longer considered part of legitimated knowledge. The objectivity of the autonomous individual required a different set of devices and strategies, more suited to the control necessary for maintaining stable representations.

Emerging capitalism, class mobility, changes in education and the devel-opment of the print medium as a primary mode of communication, all needed a quite different formulation of the topics which would enable a use of analogy that was relevant to the new social context and a broader and more varied market for print. For a number of reasons, partly related to the devaluation of rhetorical argument and its replacement by rational logic, the topics became restrictive and limited, reduced to static sets of quasi-proofs. Their apparent inflexibility contributed to their rejection by a new religious spirit that distrusted their *copia* or matter and their ability to arouse the passions, and to their rejection by the later humanists who either were those withdrawing from the social pragmatics that the topics always indicated when they were fully being used, or those who were moving into a firmly non-scholastic, vernacular preoccupation with genre and a theory of poetics which was supplanting the old rhetoric. Their failure to contribute to many growing areas of knowledge aided the development of the non-contextual and non-social logic of mathematics, which came to dominate scientific study and which may easily strip away concern for people in society.

Much more is happening that contributes to this debate, but by the incorporation in England during the 1630s of reductive versions of Ramusian rhetoric there is an educational underpinning for the denial of the topics as reasoning devices. This educational underpinning is immensely influential in downgrading the topics as argumentatively valid. During this period the 'humanities' were being located in a classical education that remained committed to a contextual study that encouraged some consideration of methodology, while largely ceasing to consider the social implications of its inquiry. Classical education, as defined by the humanists of the late Renaissance, became a halfway house between the social rhetoric of Agricola and Erasmus and the isolated rational analytics of modern science. While admitting that arguments and persuasion could not proceed by a set of fixed laws to a fact of nature, the focus on strategic manoeuvre and negotiation between examples was kept, by humanist academics such as Scaliger, well out of the social field, and often claimed access to precise truth.[46] Interest in the social, or in the immediate historical and material contingency for discussion, was by and large relegated to poetics, where the uncertainty of social relations was recast as deception, to be rehabilitated during the eighteenth century as the sublime and beautiful unknown. Furthermore, one has to recognise the monolithic nature of the classical education received by virtually all members of Western European society who wielded any power, entered any profession from the sixteenth to the twentieth century. It was, and until very recently has been, the basis for the education of the governing class and the state establishment of an English ideology. The humanities have provided for nearly four hundred years stable sets of topics for the ideological common grounds of the state.

Other stable sets of common grounds emerged within the 'laws of nature' of modern science. Bacon attempted to save a role for rhetorical topoi possibly not as valid reasoning but as illustrative and expressive.[47] The orderly qualities of the categories were their saving grace in the eyes of the new empiricists. However, even this restrictive role was perceived as inappropriate by the Port-Royal logicians who rejected the involvement in reasoning of categorical topoi.[48] The rejection is partly to do with the loss of audience for the communication of the new mathematics and sciences, the move from disputation to private inquiry, as well as the growth of mathematical logic as the source of proofs. Curiously, the categorical topoi of essence, quantity, quality, time, state and so on, were in effect retained but divorced from their social context, as the phenomenal or natural 'facts' with which science dealt. Both loss of audience and the shift from topos to fact underwrite a significant loss of the social and historical. The logical progression from hypothesis to proof to premise, is of course chronological and temporal. But people forget this process all the more easily because of the intense closure achieved by the status of mathematical 'proof'. Scientists often go on to argue that Charles' law or Boyle's law is 'true', rather than being the new scientific topoi of category, new loci or commonplaces for argument: rhetorical inventions. They often ignore that mathematics itself is based on the topoi of structure of rhetorical dispositia. Most scientists even now, as is well recognised elsewhere, have a little understanding of the topics of person and act, and ignore the effects of the observer or scientific participator on the observed or carried-out experiment.

All of these topical, non-determinist, non-self-evident bases for reasoning have, in a world after the Port-Royal logicians and the hopes of the Royal Society, a rather different role. They are not only hierarchically less important than dialectic, now called logic, but inadmissible as valid reasoning. This has far-reaching implications for their development. The 'humanities' still suffer today from accusations of being at best 'soft sciences' with no numeration, and find themselves increasingly forced to justify themselves in terms of 'proof' and 'truth', having lost much of the practice of arguing from social context and historical materiality. A consistent pattern in their attempts to justify themselves has been the tendency for each of them to claim authority for their specific common grounds, whether spiritual, ideological, psychological or social, by shifting the basis of their presence from actively agreed-upon grounds to patterns with *a priori* status: an all-encompassing God, unexamined 'natural' ideology, or corporate self-evidence. Their concern with context and methodology does not automatically imply a concern with the social and historical placing of knowledge. When the scientists of the nineteenth to twentieth centuries moved their central analogy from 'fact' to 'model', they shifted the epistemology from procedures for mapping out the world to designing it, and this shift is directly parallel to that in rhetoric between technique and strategy, between isolated device and a structural surround for methodology which encourages a reflexive engagement with the methodology of context. But the shift does not necessarily engage with the social.

TOPICAL REASONING IN HYPERTEXTS

The focus on context

During the twentieth century many movements across the humanities, such as formalism and structuralism, have attempted to address the problem of methodology by an ever-increasing attention to precise and accurate strategy, and many early computer applications in the humanities were further attempts to extend this inquiry. Earlier attempts such as symbolic logic or formalism arrived at the supposedly neutral truth-procedures of the methodologies of the sciences that dominate current studies in philosophy, and can arguably be said to have encouraged the reactive selfishness of ahistorical postmodernist aesthetics (Bernstein 1985). More recent attempts, possibly most clearly played out in the social sciences since the 1960s, have aimed to restore the socialisation of context while maintaining a critique of methodology.

Methodology does need to be studied and articulated within the humanities, but this, just as reflexivity and awareness of ethos and pathos in AI, is not enough on its own. The immediacy of social practices also needs to be restored to the humanities if studies of communication and culture are to be brought together with their value to society.[49] If awareness of methodology and context never automatically guarantees attention to social relevance, broadly speaking, the legitimation crisis surrounding political power considered by some western philosophers during the 1970s and 1980s negotiates exactly the gap between methodology however contextual, and social relevance. What is interesting is that many of these philosophers locate this negotiation in the discourse of science, partly because as to develop in Chapter 6, it has become increasingly apparent that they take for granted that the humanities are inextricably concerned with the relations between the two, and partly because the sciences do need to consider more carefully the effects of their applications in technology and how they relate to society. The rational and analytical logic of modern science has consistently decontextualised itself in an attempt to proceed on the basis of information and fact,[50] which need systems of corporately agreed logics and vocabularies to remain stable: this is the primary structure of scientific pluralism outlined by postmodernist theory that was discussed in Chapter 3. In popular perception, science appears to be searching for determinate end-limited truths. Yet the possibility of the existence of such absolute fixities denies any relevance to social context.

This separation between context and social immediacy is parallel to that perceived between political authority and value that lies at the heart of the legitimation crisis being discussed in the 1970s and 1980s. Commentators there have attempted a variety of strategies in order to reclaim the value of social materiality, and have frequently turned to science in order to elucidate their points. H.-G.Gadamer's formulation of modern scientific practice describes groups of people who form communities when they assent to convincing descriptions of reality, when they 'discover themselves in this common reality'.[51] But those 'convincing' descriptions, while based on prejudiced or taken-for-granted norms, in practice also challenge 'to a critique of prejudices' (82). 'Practice' for Gadamer, as for Aristotle, is the practice of consensual reasoning within one's community, something he refers to as 'solidarity' (87), which necessitates an education in hermeneutics, and which is vitally important to the sciences. We may not be able to question all the conventions of our community, but to legitimate the structure of that community we must be prepared to criticise those conventions as soon as they cease to be convincing.

The drawback to this particular kind of reasoning is first, that it has little way of interacting with others outside the community because, by definition, they belong to other communities assenting to different descriptions; and second, that this enclosure encourages a complacency with current conventions so that they rarely seem unconvincing. This is the main reason for Gadamer's insistence on an education in hermeneutics: only if we learn how to continually reassess common grounds and assent to new ones can the community be legitimate, bring together knowledge and value – more broadly, political action and social needs. But that insistence does not take on the difficulties that result from the structure of enclosed communities. This lacuna has been dealt with in other ways for scientific communities. Habermas suggests that consensual solidarity is tautologically self-legitimating. In effect it creates corporate systems that are internally coherent but cannot legitimate themselves in terms of other systems. For him there need to be abstract, universally convincing descriptions, otherwise there will be no way of distinguishing between a 'true' consensus that is 'good' for society, and a 'false' consensus that is merely 'successful' or convenient.[52] With these distinctions, the differences between communities are simultaneously possible to overcome in an utopian impulse to find a common set of grounds for all human beings, yet they are also markers of suspicion generating continual scepticism about the value of *any* system.

In the attempt to deal with global issues, and fundamentally concerned with the elision of consensus decisions about common grounds into corporate authority that denies participation of the individual, Habermas suggests that we hang on to universal values. The apparent contradiction here between universal values and the voice of the individual is of course the dichotomy of liberal humanism, but Habermas does argue that 'philosophy shares with the sciences a fallibilistic consciousness, in that its strong universalistic suppositions require confirmation in and interplay with empirical theories of competence' (196). Science however, unlike philosophy, does not have to 'account reflectively for its own context' (196).

Lyotard in the early 1980s devised an analysis of modern science that does away with consensus and with universals, putting in their place a 'paralogy' of systems, none any more good or successful than another. He suggests that it is not necessary to be concerned with the lack of interaction between or among systems: the condition of postmodern science is a series of discrete, concurrently existing paradigms, a paralogy that is not worried about legitimation at all. Since legitimation is to do with power, with how people give assent or authority to it, the description here, in simply assuming a position of empowerment, is inappropriate to social and political action. What Lyotard does not do is chart the dissolution of this naive arrogance of the powerful into the nostalgia, cynicism and melancholic narcissism of postmodernism: arguably the clearest outcome in the contemporary world of severing methodology from the immediacy of social practices.

Although these commentators turn to science as a place to discuss the enclosed structure of liberal humanism, or humanism in western nation-state political institutions, and although they recognise that science needs to acquire a way of involving social practices, they do not address the problem of how science is to do so, and as a result evade the larger political issues. On the whole they offer the analogical and topical reasoning of narrative, either grand metanarrative, paralogical narrative or ethnocentric narrative, as an answer. But as the experience of the humanities will attest, and as the foregoing analysis has attempted to outline, narrative is certainly contextual but it does not necessarily address social realities. Narrative is always under historical and economic pressure, ideological constraints that may be working to obscure rather than foreground particular issues.

Although science has been technologically successful and has achieved identifiable results, it has continued to remain estranged from values and ethics. As previously noted, this is at least partly as a result of technology's need to construct many different long-term markets in order to reap substantial financial profits, which has led to an extension of the rhetorical skills developed by science for inventing and maintaining a plurality of concurrent self-enclosed systems. These strategies attempt to evade social contact that might permit/allow for intrusions from the world outside which would destabilise the system. But while the systems are often consciously constructed in discrete scientific inquiry, the much larger scale of application introduced by technology and capital investment introduces patterns of knowledge that are more easily taken for granted and not consciously assessed.[53] The recent attempt by sociologists and philosophers of science to restore perspective via reflexivity, and by AI to provide environments via modelled contexts like connectionism, is a gesture to what the humanities do well.[54] But the relation of science to the rest of society can only proceed by restoring social practices to its analysis of its methodology and to its provision of context.

The humanities, inclusive of both the arts and the branches of study deriving from classical education, have since the Renaissance been the location for the study of context and its uses and relevance. Over the past two decades, computer use in the humanities has been dominated by scientific methodologies aimed at testing the truth of hypotheses. Humanities applications were thus immediately faced with a problem: that the techniques they took on were not designed to incorporate context or to encourage the socialisation of context. The concordancing, editing and database tools have provided immense advantages for extended analysis of texts and related material on publishing and printing, and a number of other fields necessary to the study of literature and the book, but they have also encouraged apparently neutral number-crunching and the lure of the definite answers that such potentially reductive activity promises.[55]

Storytelling: hypertext and the printed book

At the same time, over the past twenty years, a rather different application has been emerging: hypertext/hypermedia. This application is in the scientific community not often thought of as enhancing knowledge or logic because it is a *medium*. Yet this is precisely why the humanities are currently fascinated by its epistemological implications. Unlike earlier computer tools which concentrated on organising, defining, categorising or reducing already available written or socio-historical literary texts in order to make analysis easier and more efficient, hypermedia procedures encourage people to add to existing texts, to provide relevant context and emphasise the relational structures working within systems of knowledge. What they do not automatically do, and what there is an urgent need for the applications to attempt, is a social assessment of the methodology of the provision of context.

Hypertext not only allows for the incorporation of context but also, in this process of incorporation, produces a different kind of textuality in the same way that film did earlier this century or print in the 1500s. This new textuality offers a different methodology for representing the world to ourselves. Like any new textuality it appeared initially to offer limitless possibilities, but as people have used it, their decisions about its helpfulness or its value have recursively directed the development of the medium. Hypertext emphasises the process of learning, but only because it is new to users – there is nothing inherently expansive or limiting about its strategies; they have to be placed in social contexts. But even now, well into its development, the different textuality still seems to promise the kind of liberatory ethos offered in the parallel gesture from philosophers of science toward the apparent transcendent textuality of the arts.

Some of the most helpful hypertext products for the humanities have emerged from the Intermedia group of Brown University, where George Landow has emerged as a serious analyst of humanities computing. In *Hypertext*,[56] he, following several other commentators, takes on a post-Renaissance history for printed texts as the point of departure for his study of the empowering possibilities in hypertext and hypermedia, and focuses on a critique of objectivity as the linear, fixed and hierarchical product of print. In opposition, hypertext is presented as relative, flexible and non-hierarchical. In effect there are two arguments: the first is that print is necessarily static despite attempts such as chapter headings and indexes to render the book more flexible; and the second is that the relative is necessarily more engaging and democratic than the fixed or the linear. The first argument is pursued at length, but without much attention to the history of post-Renaissance poetics, which could be said to have developed partly as a result of people using print in a flexible way.[57] Certainly, the analysis leads Landow to claim for hypertext a 'culture more in common with that of preliterate man' (62), and an orality which is more 'participatory' and 'communal' than writing.

The second argument makes the claim that reflexiveness provides context, and that this automatically gets rid of the closure of objectivity. It is directly parallel with the claim made by sociologists and philosophers of science and criticised in the previous chapters, that reflexivity provides scientific work automatically with some kind of relevance. Landow cites the democratisation of the text that results from the ability to present multiple editions (64). However, as indicated in Chapter 2, there is no necessary democratisation involved. The user with no education in the guidelines for making editorial assessments against a background of historical and social knowledge may simply construct an autonomous, eclectic text that will reinforce private isolation. It may be significant that the analysis presented follows the frequent misreading of Derrida as privileging one medium over another, and that it refers only to Michel Foucault's early work in which he is concerned with ideological representations, not with social agency.

Nevertheless, Landow puts his finger on many of the issues that fascinate humanities users, particularly on the way that a new medium opens up different possibilities for audience and rhetor interaction. The analysis emphasises the interactiveness enabled by the network structure possible to hypertexts, and claims for the connectionist model of linking an associative use of analogy that is in itself a positive strategy because it asks the reader to do things with the text. This collocation of words – interaction, associational, analogy, network – that are significant for humanities theorists, particularly of hypertext, raises a number of questions. Associative connections may of course simply be banal, and based on self-evident assumptions that are not being considered, and may even be hidden. Furthermore, the associative 'network' may simply look like a web because of our prior expectations of a particular kind of linearity, and may turn into acceptable chains of thinking which become quite straightforward. The analytic itself is usually a form of association that has become static; a 'property' or essence is often a relation that has become accepted. What these concerns underline is the way that relativist designs are often the Janusface of determinist maps, and therefore emphasise the importance of placing any technique or device, whether it be analogy, linearity, or whatever, within a social materiality that encourages one self-consciously to assess the topics of person and act, the engagement with consensus or corporate or authoritarian stance, the position or standpoint of the reader and writer.

What the newness of hypertext offers is encouragement to sort through what the ground for its evaluation could be. That it is new to the sciences as well as the humanities is helpful since each can bring to it different approaches. Yet any discussion can only be relevant for the current historical period, and in my own case, the social context of Western Europe and North America. The position of hypertext as a textual methodology is particularly tricky for this society because one of its primary devices is to enable access to multiple contexts. For the scientist, what is new is the way hypertext inexorably contextualises methodology by insisting on its textuality as a medium. For the humanities user, what is startling is the way hypertext foregrounds the methodological strategies and assumptions in its provision of context, and provides a metacontext of immediate practice through which the context and its assumptions may be evaluated. Theory of hypertext could become a helpful site for the humanities and sciences to debate issues of legitimation and value.

With working practice, and the commentary here will describe briefly four case studies, it is possible to begin to understand that rules of thumb have to be worked out to conserve, extend or limit the potential for textuality and metacontext that hypertext offers. In practice, the seemingly inescapable focus on process and consensus can elide all too easily and without awareness into the corporate. Self-consciously watching this happen can provide a different vocabulary for sorting through the agreements we make about common grounds for argument and action.

Just as with the printed book, to which writers in the sixteenth and seventeenth centuries responded on the one hand with elaborate theories of genre that establish relations between writer and reader, and on the other with an insistence on the purity of words and the complete rejection of rhetorical and persuasive skills, so with hypertext it is necessary to generate recognisable patterns and devices for the writer and reader. One of the aspects these patterns might indicate is the degree of involvement or participation expected from the reader. Of course as a cultural skill is acquired we begin to feel competent in it if not confident. The anxious edge of consensus negotiation generates a canon of patterns, strategies and devices that are appropriate for particular social and cultural communication. Yet that canon can become expedient, unquestioned and static as popular common grounds elide into populist devices for manipulating mass media. The case studies here provide an opportunity to talk about ways of self-consciously watching and assessing that elision from consensus to the corporate to the authoritarian.

The procedure of using a hypermedia text or document is similar to writing a narrative story rather than searching for an identifiable answer. Hypertext programs vary in their detail but all provide a way of grouping together a number of texts in one place, and linking the texts to one another. Just as we might, if writing an essay on a Dryden poem, have assembled on our desk a history of the late seventeenth century, an edition of Dryden's correspondence, the diary of a well known Protestant supporter, a good etymological dictionary and a book of essays on early modern poetics; so in a hypertext we could place all these 'texts' on to a disk and as we read the poem we could refer to the other documents when necessary simply by calling them to the screen. The concept of a hypertext goes at least one stage further: that in the process of referral we are in effect constructing a context for our reading; and that if we were to make those points of reference accessible to other readers, by providing simple procedures such as an italicised word to indicate a link to another text, those readers could take part in our storytelling. Our written essay does this by citing sources, quoting other commentaries, providing footnotes, or explicitly linking our reading with another in sympathy or contradiction. The hypertext allows the writer to incorporate a much wider set of reference texts, and allows the reader to follow the critical references and links that have been made, but not necessarily all of them or in any specifically defined order. The hypertext reader has access to the context of the critic's story, but may not choose to construct exactly the same plot – indeed the reader may choose simply to wander serendipitously through the assembled texts.[58]

Let's say we want to build a simple hypertext for the Dryden poem. The poem itself will be one document in the text, as well the correspondence, the diary, the dictionary and the collection of other critical essays about that poem; perhaps also a few illustrations pertinent to the poem, or a clip from a video of a well known scholar lecturing on the rhetoric of the argument, a variety of political and religious historical documents, and so on. Once the documents have been chosen (and most programmes allow you to add more as you go on), you can start making links. If, for example, you want to get the reader of your hypertext to think about the political implications of the poem, you could set up a large number of links between the poem and the political documents you have chosen. As detailed below, that there *is* a link can be indicated by presenting the link word or phrase in italic, or bold, or red, or blue – possibly different presentations for links that are about gender and politics or links that are about art and politics. And differently presented links can be made between the poem and the other documents, either fewer of them, or coded to be presented in such a way that the reader knows they are less important. Indeed, it would be quite possible to encourage the reader to return again and again not to the poem, but to one of the political documents, so that the overwhelming experience of reading or using this hypertext essay would be to foreground the political history of which the poem is a small part. However, unlike a linear essay, the reader may choose not to follow the signposted 'important' links, and may reconstruct the context of the poem, albeit within the selected documents, by following a different path.

Story-making, or the particular re-expression of narrative pattern, is what most students, teachers and researchers in the humanities are doing in traditional methodology. Hypermedia texts can also permit a more flexible and inquiring cross-media narrative, because they make possible the inclusion of written, musical, pictorial, filmed and oral texts.[59] There is within the constraints of each programme, not only the potential for the texts to offer the commentary specific to their medium, but also to intercomment on each other in different ways. Hypermedia texts can make it possible for us not only to bring together other media but also to make another medium: hence the confusion between hypertext (assembled printed texts), hypermedia (assembled texts from different media) and hypermedia texts (whole texts made up of assembled texts from one or more different media), although most commentators now use the word 'hypertext' to refer generically to all three.

The sharp rise in interest in computers shown by humanities teachers and researchers since the availability of hypertext/media programs is possibly due at least as much to the promise of extensions to a familiar and valuable methodology as to the more approachable computer interface that the programmes offer. Fear of the machine is being replaced by the promises (and threats) of power. The apparent freedom from hierarchy, direction and consecutive linear argument, and the apparent freedom to construct our own readings, led many early commentators on hypertext to stress it as a potentially non-rational, individually enlightening structure or process.[60] Much of the initial literature has a utopian note of anti-reductive, anti-analytical triumph. Even more recently the rhetoric of the publicity for the hypermedia text XANADU indicates its romantic underpinning:

> Do you remember how you felt when you first got into computers? The feeling that everything was possible, that a whole new world was opening up for you?...Do you remember when that Dream died? But what if there were a way to bring back that dream?...THE DREAM LIVES ON....Welcome to Xanadu.[61]

But as with any strategy, the use of hypertext/hypermedia holds a lot of potential problems, many of which result from the inevitable lack of education in and application of the particular communicative devices and strategies of this new medium.[62]

Hypertexts may be *relative*, but they are relative only to the person who chooses the contents. One of the most pressing methodological problems is that although context is made available, and the user is far more readily able to go out into it within hyperspace and further their understanding, that context is specifically selected and inevitably biased.[63] The very 'friendliness' of hypermedia programmes is a problem. Their procedures often conform to our ideological notions of representation and critical inquiry, and hence appear trustworthy like the directly analogous procedures of critical and theoretical commentaries and textbooks. Yet unlike the latter traditional procedures, the humanities community is not yet properly aware of the current limitations and extensions of hypermedia, which are as often rejected by non-users through fear of their obscured control as users euphorically accept them.

Hypertexts are never *non-hierarchical*. What is frequently obscured is that the hypermedia text is composed of a number of smaller pieces of various texts – whose presence has been chosen. Furthermore, these texts are usually linked together in order to guide or encourage the user to contextualise, yet these links which are both implicitly significant choices and hierarchically organised through the action of the link itself, are not foregrounded.[64] For example GUIDE, the programme with which I had most of my early experience, had three main ways for linking text. The first was 'note', which provides a temporary link to another piece of text that appears on the screen for as long as you depress a mouse button. The second was 'replacement', which replaces a section of the current text with other text, from which you may return to the initial text or move on to other replacements or 'references'. The 'reference' was the third kind of link which replaces one text with another. You may return to previous texts or you may wander on through others following reference links. The point here, though, is that each kind of link provides a different kind of 'level of importance': a note is less substantial than a reference. Certainly, it may not be as helpful or effective in the end, but there is an implicit hierarchy of information coded by the syntax of the program.

Hypertext may be *flexible*, but it has to have structure: more or less openly indicated. A related problem that occurs precisely because the medium is relatively new, is that any user may not only be undereducated in the communicative devices of hypermedia, but may frequently also experience intense anxiety for the reassurances of the printed book. There is a desire for a fixed text that can be checked back on for a multitude of purposes:[65] and for a fixed text not only in the sense of one that uses conventional and highly learned strategies, but also in the sense of an unchanging printed product that can be matched up against precise wording. This is partly to do with the different cognitive response called for: reading books involves a physical activity that the reading of hypermedia texts makes virtual. But this is less of a radical shift for humanities users who overtly use virtual strategies in many of their reading skills. Yet people still want to know 'where they've been', or 'what page' something is on: recall the way we often remember the location of a phrase in terms of 'half-way down a page' or 'at the top right-hand, about one-third of the way in'. In response, many programs began to offer a choice of pagination or no pagination for previously continuous scrolling of texts, in other words the screen-full of text is numbered. Other programs actually attempt a schematic representation of books by including a central divider, margins, gutter margins, and an increasingly shaded left-hand side of the screen (and decreasingly shaded right-hand) to indicate from the depth of the pages 'how far' you've read.

More commonly, most programs now offer schematic maps of 'where you've been', and permit you to backtrack and choose alternative pathways. For example, the program HYPERDOC[66] allowed you to press a specified function key and instantly be shown on the screen an iconic representation of the last seven or eight documents you had looked at. Graphics, text and photograph are each designated by an iconic shape such as square, diamond or circle; and if you move the mouse pointer over a specific icon and depress one button, a brief summary of the contents appears directly on the screen. Pressing the alternative button will bring the full document up on the screen and you may continue to wander. The IMP software that was used for the fourth case study has a permanent column on the right-hand margin of the screen to indicate with small quasi-facsimile icons the preceding ten texts or illustrations you have viewed, any one of which you may return to by using the mouse pointer.

Another fear that emerges from the loss of the reassurance of the printed book is fear of size. The hypermedia documents produced so far have tended to be large, and users are frequently frightened off by an assumption that they have to read the whole thing, as they might a book – indeed that they might have missed something. The reader who conducts library research gets used to surveying the stacks for a serendipitously related work, but at least they can 'see' what's on offer and have learned rules of thumb for choosing. Hypermedia structures often hide the totality of available texts. A similar anxiety is induced in readers who have to move from open library stack systems to libraries with closed stacks accessible only by catalogue. Just as with time we get used to the catalogues, sometimes finding them more beneficial in the end, just so I suspect that increased use of hypermedia procedures will offer its own benefits.

Underlying the fear of size, and possibly the most important area of potential for obscuring structure, is the difficulty of assessing the inbuilt biases of large hypermedia documents. The apparent range of contextual apparatus that such a text can provide may, rather than raise consciousness of context, instead yield systems of self-validating context. The selected texts may simply provide a tautological support for each other, as in the construction of the multi-document newspaper or magazine in which advertisements reinforce copy which in turn reinforces editorial.[67] Hypermedia, and periodical publications that have emerged from the printing presses since the seventeenth century, have much in common as communicative discourses. Related to this third problem is the last one I want to raise here, and it is in many ways the most important because it indicates the place where the social needs to intervene: that just because hypermedia strategies and structures may parallel current activities in humanities study they may foster uncritical nostalgia for, rather than helpful commentary on, existing methodology.

However, while hypertexts are not unproblematically relative, non-hierarchical or flexible, the obverse face to these difficulties has immense potential. Both the textuality and the metacontexts insisted upon by social and historical contextualisation belong to the craft of articulating rhetorical stance. What is thought to be inarticulable in many areas of knowledge and expertise is in effect a failure to remember society. If we attempt to be conscious of the drawbacks, we can use them to foreground and provide a location for the discussion of methodologies of context, precisely because the hypermedia text exposes the problems so clearly. There is a concurrent need for education in the medium and for development of ways of using it constructively: these activities are not the same and cannot be fully simultaneous. For five hundred years Western Europe and its colonised outcrops have been using, responding to and learning about the printed page. Our education systems in the West by and large necessarily focus for at least ten impressionable (pardon the pun) years on teaching the young how to read and write in many different genres. Nearly all our knowledge is conveyed by way of print, and certainly most people in the West have to survive in a print society that demands that they at least need to know how to negotiate driving licence applications, telephone directories and electricity bills. But with hypermedia we are all very much in the same position as the fifteenth-century reader of print: by definition we are economically privileged, somewhat educated in the medium but tied inexorably to an earlier mode of communication, and in the position of both learning about and developing that medium. The medium will not be self-revealing: it will be what we make it into, and we need to exchange accounts of our experiences to clarify the advantages and limitations of hypermedia. The following accounts are put forward as case studies which need critique. That said, it seems appropriate to outline a number of approaches to hypermedia application as examples of my own learning curve.

Hypertext narratives: case studies

Hypermedia projects of which I have had experience can be grouped under four approaches: topic-driven hypermedia texts, central text hypermedia, multidocument hypermedia and hypermedia nests.

The topic-driven hypermedia text

One of the first approaches I took was the topic-driven hypermedia text, and the choice may be significant. The Chemical Pathology Department of St James' Hospital in Leeds wanted to develop a hypertext to ease access to current information and to encourage helpful expansion and combination of the information.[68] Although the project began by hoping to make possible undirected and unhierarchical wandering through the system, it was almost immediately realised that the mere pre-*existence* of the information to be put into the hypertext meant that substantial categorising and hierarchising had already taken place, which could not be disrupted without unhelpfully disordering the expectations of users. The known information had already been organised by a highly professional body (the medical academy) which trains its members for seven to twelve years in exactly those methods of ordering that we had been proposing to untie. Furthermore, the definite practical end for the main user might well be a decision radically bearing upon the health of a living human being. Evaluation of options at speed would be essential, and any dismantling of expected order might be a drawback.

The result was a hypertext within which the user could effectively follow up connections between pieces of information surrounding a specific topic. For example, a query may have come in regarding paracetamol poisoning in a child. The user could begin with 'diabetes', 'renal failure', 'paediatrics' or other relevant topical vocabulary, and be able to wander quickly around the available relevant pieces of information including graphics such as a chart of progressive paracetamol poisoning, to get an idea of whom to call or indeed what to do. The key issue was to ensure that those people who linked the 'relevant' texts together while the hypertext was being constructed, also had expectations similar to those of the user or at least not fundamentally different. This not only aids the user in a pragmatic sense, but also makes it easier to incorporate the user's suggestions, recorded on a notepad attached to each document, for links into the text. A similar structure seems to have emerged from the Brown University BioMed project which describes the biochemical pathway system. Here there is a vast amount of material which the undergraduate student is encouraged to explore, but that student need not look at the entire system. They can choose to look at a small unit without any pressure to a total or complete overview or to a need for contextual validation. This process of topical specialisation is well suited to areas of study in which, once the ground rules are known, the student's job is to learn how to play by those rules. Such hypertexts are set up by highly trained people for educated specialist users, to better present the formal, corporately held directions of information and make more effective the sense of a necessary answer or conclusion.

Topic-driven hypertexts are less problematic to construct than many other applications because by definition they count on a stable approach to material that does not unduly disrupt the expected order. In many ways they are more flexible and speedy enumerative bibliographic systems, and have no necessary connection to any social materiality that would require their users to engage with and assess the choice of material or the kinds of links that had been established. But because they effectively provide a multidimensional filing system, they do make possible far more combinations and links between existing pieces of information. This *may* loosen up the specialist use of that information, but it is not a fundamentally different use of the medium or a shift in approach to medical discipline. When the printing press emerged at the end of the fifteenth century, the practice of keeping 'commonplace books' became more widely adopted in Europe. Whereas their medieval, manuscript-based counterparts would write extensive commentaries in the margins of books next to particular passages, the children and adults of the sixteenth century would keep commonplace books to write down important quotations or significant phrases and stories.[69] But not until the latter half of the century, when the practice yielded up the topically focused essay form in the writings of Montaigne and others, did a radically different use of the medium emerge. It may take as long for a new approach to topic to emerge from hypertext.

Central text hypermedia

Central text hypermedia, my second offered approach, is also based on known, more or less practised rules for research, particularly in textual editing for the humanities. This approach can provide either encyclopaedic information or an informational shell surrounding the central text, a text which may be a person/writer or a literary artefact. Again, Brown University early on developed an encyclopaedic central text called Intertext, which revolves around important authors and their works. Say we want to look at Charles Dickens' *Oliver Twist*: such a system could begin with Dickens and provide biographical, bibliographical and other contextual material for the person, as well as a very wide variety of contextual information on the novel itself, which might well interlink with his journalism or with other similar texts produced at the time. Students should already be carrying out this kind of contextual study, but an encyclopaedic hypertext makes it quicker, possibly offering the student a wider range of issues than they might themselves have chosen. Most important, the contextualisation that the reader is doing when they use this hypertext becomes part of the act of reading: much more immediately and pragmatically than usual, context becomes part of the text.

When the researchers who had been working on the St James' hypertext and myself turned to building a hypertext for a course in English literature, we were immediately seduced into the possibilities of an encyclopaedic form.[70] The lure of encyclopaedic form is one of the dreams of Frankenstein. Let's say you start with a short story. Usually you have one hour with a class, but with a hypertext you think 'Now I have the opportunity to fill out all those areas we don't have time for'. So you begin with an important piece of criticism, or with a detailed language study, or with a parody of the text, or with a comparative contemporary piece, or with a note on the editions, or with a publishing/reception history, or with historical background (all of it!), or with censorship, or with cultural context, or with semiotic analysis, or with the film version, or with.... It is like trying to build a comprehensive library. It just doesn't stop if you start this way and starting this way can make us manic. Helpfully for once, lack of money and time made it impossible for us to fulfil the dream. We rationalised that students could do that kind of study in a library or in a resource centre in the real world, so why should we attempt to do it for them?

What emerged instead was a slightly different approach, still based on the central text, but aiming to create for it a shell of critical approaches and practices rather than an encyclopaedic environment. Although students can do this kind of background contextual study, they often do not know how, so we decided to produce a hypertext which placed emphasis on the research and learning strategy itself. Instead of using context to 'firm up' a reading of a text, we wanted them to learn about the structuring that surrounds that text and its available readings. In a new or different utopian mode: we were after a virtual, not an essential central text. In other words, we wanted not a central text with surrounding material to back it up, but a central text with context that enabled different possible readings. This concept will be familiar to computer users of 'virtual' machines, but is different in that unlike with the machines, you cannot do the same things with both virtual and essential texts. For example, if the words on the printed page make up the essential text of *Persuasion*, then the socio-historical, cultural and critical structures, the economics of early nineteenth-century England, the position of women, the emerging role of the publisher, and so on, make up the virtual text of the novel. Rephrased this way, the difference will be familiar to those who teach literature, since it parallels the difference between study of a text as a verbal icon and as a cultural icon. Either way the potential is there to study the text from a position grounded in current and assessed social practices, but the virtual text will not do so automatically, any more than the essential.

Using GUIDE we constructed a hypertext shell to be used for the study of Jane Austen's writing. In the process we decided that some form of hierarchy was indeed helpful to the user, although no more than three layers deep. Recognition of the importance of hierarchy came with the recognition that we were teaching, and that teaching is always implicitly if not explicitly a mode of guidance toward values we wish to encourage. This guidance does not signify that a set of standards has to be inculcated. Rather the guidance is intended to indicate that value derives from being conscious of the basis for our choice of actions. Value is not ideal, it is practical – hence guidance is partial at best. But one aspect of guidance is about the techniques and strategies of becoming conscious and retaining awareness, and this is where a clearly delineated hierarchical structure is a useful technique for reminding ourselves which areas or kinds of texts we would find helpful to focus upon. Further, we discovered that just like the St James' project, we were subject to disciplinary procedures and codes that seemed inexorably to impose a direction on the free wandering of the user. Simply to aid in the construction of the hypertext, seven main areas were proposed – author, genre, critical texts, book as object, language, primary texts, and history – but these had no separate status for the user. This was an important misunderstanding of the user's needs; it was not yet appreciated by us how daunting an experience is this apparent free-wandering in hypertext. But more significantly, we had not yet appreciated how the foregrounding or flagging of our own structures was a valuable aid in alerting the user to our biases and to our principles for guidance.

This concurrent but not simultaneous realisation is a clear example of the difficult tension between the two activities of constructing the medium and learning about it at the same time. Indeed, bias and predisposition emerged unexpectedly all the way along. The most startling was to find that, silently, the hypertext turned and returned the user to the history section whenever a relevant choice of link was offered. When I say the hypertext did this, I mean of course that we were doing it, but until this moment I had not realised how significant history was to my own pedagogical techniques. This particular project was still over-ambitious, too large, and based on an early and rather cumbersome hypertext programme. But although it failed, I still think that the 'shell' approach offers a potentially useful technique for teachers who want to displace the essential text and stress the cultural and critical structures of humanities discourse that make that text possible.

Multidocument hypermedia

A third project with which I was involved in a peripheral way, was the Hartlib Papers Project at the University of Sheffield.[71] It is a clear example of a multidocument archive that uses hypertext as a presentation device, what I call multidocument hypermedia, and which is close to the structure of a periodical publication. The Hartlib papers consist of thousands of letters written to Samuel Hartlib by a large number of people between 1630 and 1662. Reading through them gives one the unusual experience of reading an immense magazine from a period in which magazines were not yet current, and underlines the affinity that hypertext has with the modes of layout and reading techniques in the printed periodical press. The key aspects to the use of hypertext with the Hartlib papers were first, that none of the ground rules for the co-presence of the documents can be assumed, and second, that the documents themselves are preselected. All the letters are together in the collection because they passed through the hands of Samuel Hartlib, but why they cover the diverse topics ranging from optics to the colonies that they do, what the epistemological set of the group of people participating was, is difficult to recover. Interesting concurrent developments in learning and constructing knowledge arose from our radical lack of understanding about the seventeenth century, exactly which lack can make hypertext a valuable medium for presentation and can turn it into a research device.

We might have been tempted at first glance to impose a set of widely accepted rules for seventeenth-century knowledge upon the material, and this is in effect necessary for twentieth-century readers even to begin to read the archive; but then it became clear to the main editorial team that certain assumptions did not suffice, were contradicted or challenged by the material. Links could be made that we might suppose would go somewhere, but suddenly they were resisted, indicating unexpected points of significance that needed new constructions and approaches based on experience with the material. To take just one example, the metaphor of the beehive for society is something with which we are familiar. However, the profound concern of nearly every major contributor to the collection with the community organisation of bees and the visual designs and diagrams for the architectural construction of homes adequate to that community, was quite unexpected and needed radical assessment of the connections which were to be inserted into the hypertext.

The vocabulary used to describe this process of learning and assessment could be directly parallel with scientific observation and interaction with the phenomenal world, with its stress upon modelling and the significance of failure.[72] Where it must differ for a hypermedium is in its process, which does not have to conform to the logic of rational analytics, nor to pragmatic empiricism, nor to mathematics. Rather the process is once more analogous to storytelling, finding a good fit between conventions of communication and material experience, and the use of hypertext emphasises those narrative strategies. The hypermedium can alert the constructor to epistemological breaks and rifts which can then form a basis for new approaches to the material. But at the same time, and only as helpful as the constructor's work, it can encourage the *user* of the hypertext to think differently about the seventeenth century as the links are followed. Indeed, the tension between learning from and constructing the hypertext is clearly delineated and even exacerbated in multidocumented hypermedia texts, for as often as the construction selects a potentially enabling series of links and connections it directs the user away from other selections: we might choose to link astrology with outmoded scientific practice, and forget the links between astrology and contemporary ecology. There is constantly the uneasy awareness that many of the material's significances will not even be noticed simply because we cannot recognise the signs of its different resistance.

Hypermedia nests

The final project which I offer for critique is the currently on-going work with Margaret Beetham at Manchester Metropolitan University, which began as attempts to build a set of hypermedia 'nests' as aids for teaching late nineteenth-century material on the periodical press in Britain. The topic has been chosen partly because periodicals share several aspects of discourse strategy with hypermedia. This, and a conviction that hypermedia applications at this stage must include procedures for foregrounding their strategies and structures, defines the emphasis of my critique. In an immediate and unusually defined way hypertext/media raise many issues already raised by periodical publications. However, the common technology of print, and the critical and theoretical focus on the single-authored book as the primary legitimate printed product for writing, have obscured the very different conditions surrounding the periodical.[73] Both periodical publications and hypermedia texts communicate by way of chosen collections of smaller texts, related in various foregrounded ways by design and layout. More enabling for extending the reader's contextual responses, neither medium has a single writer nor a single text, neither insists on a specific order of reading, and both can and often do offer a mixture of media – the periodical has photograph, headline, cartoon, printed text, among others.

Both media raise queries about the romantic framework of literary criticism which focuses on the 'author' as an original genius, on a consistent text and on an unproblematic medium.[74] This framework forgets a number of contexts, respectively: the social basis for the writer and reader, the technological and economic effects on writing by way of printing and publishing and audience relationships, and the epistemological implications of different media. Much cultural and theoretical work is attempting to restore these contexts to the broader study of language, literature and communication. With possible significance, the extensions frequently involve consideration of different media and stress the variety of response that is generated, and hypermedia texts, as suggested earlier, make it possible not only to bring together other media but also to make another medium with a different textuality. Critical consciousness of the effects of the periodical press has been non-existent until fairly recently, probably just because they do not fit the romantic framework. One result is that critics were relatively slow to appreciate the limiting and enabling devices of the medium. The development of hypermedia needs to address this issue for itself, and one way of proceeding is to ensure that hypermedia texts are on a scale small enough to encourage the detailed studies of particular gatherings of context that are necessary to the self-reflexive and conscious explora-tion/appreciation of methodology appropriate to this stage of learning about the medium.

In practice the hypermedia nest project with periodicals is aiming to produce a series of very small units. Initially we suggested only three different periodicals, and only one issue of each to provide a focus. Around these three issues, there were to be only five or six associated topics such as print technology, writers and editors, illustration, serialised fiction, and some others. Each topic would have at most ten to twelve texts, of various media, relevant to it; each topic would also be accompanied by an outline of the links we incorporated into the text and suggest further resources. Even with this minimal context, we would already have 3 (magazine issues) × 5 (topics) × 10 (texts), or 150 texts. But since this would not be an unreasonable number of extracts in a background textbook, we hoped it would not be too large. In the event, the size appears to be satisfactory for a one-hour tutorial session, but to ensure the possibility of generating even mildly sophisticated contexts, we have focused on only one issue for the first nest.

Apart from including a number of devices to allay book-anxiety, the emphasis in the programme is on the making of links between texts. There are a variety of ways that this can be carried out. First, the teacher may choose simply to refer the student to the hypermedia text in a completely uninvolved way. The links made will therefore only be those that come with the construction of the text, as outlined in accompanying printed material. Second, the teacher may choose to add a number of links they consider helpful to the student of their particular course. Third, the student may be encouraged to use a 'notepad' facility for suggesting links that the teacher, possibly after discussion with a class, may then add. Fourth, the student may be encouraged to make links directly.[75]

At the simplest level the teacher and/or student may be looking at just one topic within one magazine issue. If the creation of links is encouraged, the texts around that topic will gradually acquire a web of connections, weaving a contextual nest around the topic. The users may choose to keep the nest individual, with each student having a private nest; they may choose to develop it under discussion, or consensually; they may choose to develop it as a group, or corporately. A particularly enabling set of links may be used as the basis for other individuals or classes, or it may be swept away. At a rather more complex level, these topical nests could be interlinked with others, not only those around one issue of a magazine, but also those around the other issues. Indeed, a course could aim to do just this: create a contextual approach through hypermedia construction on a large scale. While potentially valuable, I would suggest that this approach at this stage of our understanding of the medium is seductively totalising. It introduces context, yes, but the completeness of systematic coherence that it promises may forget the need to socialise that context, to make it relevant to our daily lives. The main point about hypermedia nests is that they are on a scale small enough to encourage assessment of the implications of contextual strategy or patterns of links, on social stance and actions. There is nothing intrinsically social about their activity, but their construction may make it easier to assess the social.

Rhetorical implications of the hypermedia medium

Strategies, no matter within what context, are not intrinsically socialised, and with the massive development of hypermedia at the moment there is an urgent need to make decisions about the implications of its methodology.[76] Just as print got itself into a position where the book became the valorised cultural object and we are only now beginning to reassess the activity of the periodical, so hypermedia needs to address the dangers and seductions of a multitude of potential directions – not the least the nightmare of Borges' universal library transferred to the living room.[77]

For example, the concept of hypermedia nests faces us with an immediate methodological problem which I have no doubt will have faced other writers of hypertexts: is the nest of significance constructed by a web of consensual choices effectively different from the island of meaning constructed within the pluralities of science by systematic selection based on corporate policy? To return briefly to the legitimation crisis and the questions it raises about power and knowledge: is the textuality of hypermedia self-enclosed and self-legitimating, or can it look outward toward society for assessment of its value? Does the social context for its apparently consensual methodology lead toward the bland sameness of corporatism or toward difference or toward both?

The consensual web starts out as a consciously admitted construction, agreed by a specific group of people to answer their particular needs. It addresses the immediate social context of those people who are trying to sort out the most appropriate stance for action within it or from it or about it. The corporate system in many ways is often a result of attempting to extend the consciously constructed position, which may have worked very well indeed for a particular moment, past its immediate context. The strategies it initially employed are used by the corporate to address another set of events, and if they are helpful are used again and again, eventually becoming taken-for-granted ground rules. The approach is systematic because the results of each action derive from the same basic strategies and thus have a set of shared common grounds that allow them to cohere into a set of relationships. The approach is corporate because it does not insist on the constant renegotiation of consensual agreement, but acts from a set of taken-for-granted common grounds without asking for a reaffirmation of the grounds. While corporate systems are pragmatically important, given the apparent impossibility of constant renegotiation, they are also on their way to fixed sets of ground rules that result in reductive, non-contextual and non-social applications. Yet this elision of the methodologies of contextual reasoning from consensual agreement through corporate system to reductive totalities is usually too diffuse and massive to assess in terms of social needs: one cannot generalise, for at times the consensual may be appropriate, at other times the systematic.

The further along the line away from social immediacy and from consensual contextual reasoning, the more probable it is for the systems to become totalising sets of self-enclosed worlds, as in the representation of mathematics. The humanities have provided a site for the study of the interaction between these methodological approaches as enacted in the print culture of the West since the Renaissance. Each area of the humanities has developed skills in assessing the appropriate point to which strategies may be extended without too sharply distorting the social contexts in front of them; and each area has developed guidelines about sets of ground rules and where and how they can be broken, shifted, transgressed. In literature in a print medium, this is precisely the attention of rhetoric, poetic, logic and grammar, and their activity within printed books and other printed texts has been elaborated and sophisticated for five hundred years – to the point where we take much for granted.

The development of hypermedia texts which at the moment allow for analogical and contextual reasoning, as print encouraged the analytical and rational, can help the humanities areas to articulate more clearly the skills, guidelines and methodologies they have been developing. And the medium also offers, partly just because it is so new, the potential to encourage an assessment of the social implications. Sciences, which take their contextual basis largely for granted, will be able to benefit even within their own current terms from the articulation of contextual methodology; but also, the current crisis they face resulting from their historical severance from and ignoring of social activity, may well be offered a point of engagement in the potential for the socialising of contextual reasoning that the humanities can pursue with skill and enthusiasm in developing the hypermedia. On the other hand, with this powerful tool for contextualisation in their hands, the humanities may be lured into descriptive or explanatory systems rather than consideration of social needs. The primary way of avoiding another reductive tyranny analogous to that which the printed book synergistically exacerbated in modern science, is to remain aware and constantly attempt to become conscious of the rhetorical implications of the hypermedia medium. If hypermedia encourages the humanities to articulate guidelines for contextual reasoning, then questions about application and assessment can be extended from the categorical success/failure divide to broader issues of social value. We cannot predict the effects of this medium, but we can choose directions for it and engage with the developments that emerge.

The sensory delight of science, its aesthetic joy and games-playing pleasures, are elements of textuality and of material practice, with social and ethical effects, all of which science obscures from its public and keeps tacit. The long tradition of impeccable performance, and fundamental distrust of language and writing that is seen as inadequate, has lent to scientific research the mask of Aristotelian rhetorical structure, the introduction or proposal, method, observation, and discussion/conclusion, while depriving it of persuasion, social location and historical agency. I have argued that computing and its theoretical counterpart AI, in their exploration of knowledge representation and expert knowledge, do not understand the textual work of the natural sciences in their interaction with the world, and that this misunderstanding has frequently controlled the development of computing as a medium. Yet I would go further and suggest that computing has its own textuality which it too is obscuring from the public, and to no good purpose.

The complex of technology, industry and business, needs the textuality of science to remain hidden, in order to maximise repeatability. But computing is not an essentially capitalist technique, even though technology under capitalism has extended its potential. My own experiences working with computers, which began when paper cards were still being used in the 1960s, were overshadowed from the start by the inability and unwillingness of computing science to articulate and value its own textuality. The journey made in these chapters, from textbooks to theory to practice, so that they permeate each other into a particular epistemology, is one that has tried to find an engagement that lets people communicate better to other women and men through computing. Trying to compete with the computer, trying to control it, is only part of working with it as the focus for a set of craft skills belonging to some human beings, that encourage us to engage in different ways with other human beings. Hypertext is currently helpful, particularly for arts and humanities users, because it mimics a kind of engaged textuality that we recognise in our more familiar world. But the textuality of hypertext, the software and hardware structures that allow that performance medium, have their own engaged textuality which only the computer scientist can fully appreciate. Transparency, seeing through all the layers, used to be a keynote. But for some time now it has been impossible for any one person, even a computer scientist, to see all the way through. Perhaps this is how the invisible membrane comes into action.

We can assume that the development of hypertext was an effort to use the medium to speak to other people, but unless computer scientists themselves take on the responsibility to articulate rather than keep tacit the social and historical effects of their medium, computing science, like research science in the natural and hard sciences, will simply develop its own glass womb, bringing into birth technologies for those exact reproductions on which technoscience and industry depend. Computing science, no more than science, needs to be this way. There is no reason other than the structure of capitalist nation-state ideology why it should not develop more closely as a medium to be used between and among human beings for their better understanding of each other, rather than as yet another medium for repression. Rejecting textuality does not leave neutrality but oppression: it deprives people of voice.

5 Feminist critiques of science

From standpoint to rhetorical stance

RECAPITULATION

The commentaries on rhetoric from the writings of Aristotle and Plato that have woven through these first four chapters have had two nodes of counterpoint. First, the strategies of power, knowledge and textuality that inform modern nation states are discussed in the monolithic classical sources and used by, or to which many present-day commentators return, in their analyses of politics, society and representation, as well as their sometimes concurrent analyses of science, artificial intelligence and/or their applications. Interpretations from the classical tradition infuse the structure of these critiques with concepts of idealism and universalism and their concomitant concepts of representation, transcendent beauty and plural replicability, whether or not they are upheld or dismissed. In a sense this is to state a commonplace, but it also pursues the extraordinarily broad extent of the traditional ground in contemporary thought: that philosophical and political analysis begins with and develops from, but is inexorably linked to a tradition that returns to, classical Greece and the first records of such an approach. But a node of counterpoint emphasises that the strategies of power, knowledge and textuality found in classical texts are based on material conditions different in kind from those of today: we have now an exponentially growing population for whom capitalism and nation has made representation a useful strategy, yet for whom direct democracy is impossible;[1] a constituency whose enfranchisement heightens the contradiction; and different technologies radically changing the negotiable and probable limits of not only the oral and the written but also of all communication. What I develop in this chapter is a third node, one that suggests that by bringing a broader approach to rhetoric, which is in any event a way of situating communication, together with the concept of situated knowledge, we can begin to understand the importance of textuality for current studies of epistemology. The appropriate term that I offer is situated textuality.

As if oblivious to the political and social materiality of history, many critiques of the rhetoric of politics and science use the classical references as though these radical changes had not occurred. They also entirely ignore the commentaries on the rhetoric of textuality as it moves from the oral to the written, commentaries which offer startling insights into the parallel movement of the verbal to the printed, and then to the variety of media that are today produced. While some critiques still posit idealism, in the general rush to dismantle reductive objectivity many now fuel the sense that 'context is all'. What in effect happens is a shift from ideology which sustains idealist notions of beauty and truth, and from the transcendence of inadequate language by represented and determined subjects, toward discourse which negotiates along the axis of ideological representations: choosing contextual nodes on which to explore and elaborate, yet remaining systematic and inexorably tied via that axis to the ghost of idealism, universalism and an inadequate language.

As elaborated in Chapter 1, ideological representations, subject positions and the negotiations along the axis they construct, that are discussed by psychoanalysis and discourse studies, are dealing with the ethos of the capitalist nation state in the West, particularly those states with the representative democracy of a liberal social contract. The history and theory of politics in western universities has demonstrated that although various,[2] the representations and positions constituted slowly over the past 300 to 400 years, in effect protect and preserve the profit-making techniques of a small minority. Those representations and positions have been exclusive of the poor, of children, of unpropertied men and of all women. They are also, therefore, inextricably exclusive of racial difference and differently abled people through the imbrication of race and bodily ability with class. As parts of the excluded population began, technically, to be permitted citizenship and 'subjecthood', many representations were already in play. Indeed, subject representations may well have achieved such rhetorical effectiveness precisely because they control the entry of previously unpropertied men into relations of ruling power, which they began to do in England in the sixteenth and seventeenth centuries as the strategy of capitalism permitted them to accumulate profit and property. For those who entered citizenship chronologically later, the range of positions available, for example to women, was necessarily restricted. Not only were these 'new' positions often predicated on a recognition of subjecthood only if women represented themselves *as* men, but also there had been many tacit positions allocated to women – unrepresentative but embodying – so that when they became 'subjects' the invisible embodiments became visible representations and ways had to be found to value them: hence 'medals for mothers'; hence the welfare state. The welfare state could be thought of as a way of containing the suddenly released-into-representation embodiments of women and to some extent the excluded condition of the working classes.

Working with the ideology-subject axis directly and with the discourse studies that analyse its embodiments, is necessary if we are to understand the oppression and repression under which and in which we live. Yet taken alone, this work can deprive one of a sense of agency, predicated as it is on the self-defining and determining systems of ideology.[3] If, however, one looks at the work done by those outside ideological representation on articulating identity or entity, if one looks beyond relations of ruling power to the much messier areas of living than those addressed by ruling power, then, while we are not completely leaving the systematic behind, for it is always contingent, we are working in areas of agency: areas in which individuals and communities relate to society and each other in ways other than through ideologically representative positions.

Work on understanding lives and practices beyond the ideological goes on in many places, for example in some postcolonial theory and commentary. Of particular importance to science and computing, and I would argue the arts and humanities, is the standpoint theory elaborated by the feminist critique of science. The remaining two chapters will discuss the ways that standpoint theory has opened out our understanding of science and power and science and knowledge. This chapter will study the way that standpoint theory also gestures to the arts to complete the articulation or representation of the excluded and assumed tacit knowledges in science. The following and final chapter will offer a critique of the lack of any sustained standpoint theory in the arts, and will discuss two (among many) aspects that may extend the debate: first, that the canonically valued arts and mainstream aesthetics are just as privileged and socially exclusionary as the sciences, working to keep profit to a few in the media industries, and that this severely compromises the gesture from science. Second, that the concept of tacit knowledge in the arts is different from that in the sciences, and that the different kinds of tacit knowledge invoked are parallel to the different kinds of exclusion from the relations of ruling power with which standpoint is concerned. The reason for this exploration is to work on the possibility that if the different textualities that result from these different concepts of tacit knowledge can be situated within social practices, then perhaps a rhetoric of situated texuality may indeed offer something to the feminist critique of science. These two chapters will outline the feminist standpoint critique of science, apply it to the arts, and see how this affects concepts of knowledge and textuality.

Feminist critique of science

Feminist critiques of science are split between the standpoint and the postmodernist, and many include attempts to find common ground between the two.[4] This is frequently problematic since the 'postmodern' is usually seen as relativist and standpoint as a pragmatic empiricism often lacking the flexibility of practice. Mainstream postmodern accounts of science such as those by Rorty or Lyotard can be criticised as profoundly ahistorical, but feminist postmodernism comes to science with the historical postmodernism of the postcolonial turn.[5] It questions the power of science at any one historical location rather than contemplating with melancholic cynicism and wistful bewilderment the actual disempowerment of the supposedly empowered scientific intellectual. Hence one finds analyses such as S. Hekman's *Gender and Knowledge*[6] in stark contrast to the descriptive work of Woolgar and Latour.[7] Nevertheless, without that immediate experience that roots the postcolonial in an analysis of specific positions of ideology and discourse, the approaches of feminist postmodernism to science can be brittle. Too often they reveal the relativist underbelly for the soft skin of positivism that it is, mimicking the anarchic/authoritarian dyad in nation-state politics, and the arbitrary/absolute dyad of post-Cartesian western philosophy.

Standpoint theory, however cast as empiricism, is not positivism. Positivism is largely defined by the representations of ideology within institutional space. It is specific to the autonomous individuals of the liberal democratic social contract and is the pair to postmodernism. Postmodernist attempts to pair itself with standpoint involve an excruciating tension as the strategies of the empowered turn toward the excluded; it is a colonising turn, implicitly self-justificatory. In contrast, standpoint comes from relations of non-ruling power: it is about the knowledge that is structured by experience of people working outside of systematic subjectivities. In answer to the immediate question: how can you/we escape ruling power? – we cannot. Yet we are often excluded from it and we do not simply cease to act. Aboriginal environmental science, traditional African science, women's science in the sixteenth and seventeenth centuries, the scientific knowledge of the 'wives of scientists', of 'women in the laboratory', of lab technicians, are a few examples of excluded scientific knowledge that have been brought into contemporary consciousness. Despite the recognition by postmodernists that science is an institution, it focuses on descriptive and analytical considerations of the club culture of science. Standpoint tends to be more explicit about the wider relations of power, from technoscience to the domestic ecologists, so that its approaches are more critical and revisioning.

Rather than attempt what I think of as a flawed conjunction of post-modernism with standpoint, this discussion will explore the way that standpoint approaches to scientific knowledge and power share common ground with approaches from contemporary studies of the history of rhetoric that consider the way that knowledge acquires different kinds of power through different kinds of legitimating practices. It is no surprise that standpoint theory and the history of rhetoric comment so helpfully on each other, since standpoint is also a part, a valuable and difficult part because it is non-systematic, of the legacy of Marxism. Marxism offers an historical response in Western European countries to the loss of rhetoric[8] as a moral and ethical training in strategies for agency and empowerment, a loss increasing from the seventeenth century and nearly complete by the nineteenth. It reinstitutes Agricolan dialectic 250 years after Ramus and Scaliger nailed it down to a reductive syllogistic, but reinstitutes it in the knowledge of a disempowered constituency called the working class. By neglecting issues of gender, Marx himself maintained the contradictions of the capitalist nation state and did not develop a critique that could see beyond ideology or discourse and thereby also enable/empower people of different races, abilities and ethnicities. However, the fact of seeing an unseen, and writing in its implications, of making visible the embodied but not represented constituency of the working-class man, was a material act of a rhetorical stance that overturned the obliterating strategies of the social contract, asked people to remember to re-member, to take the embodied back and to work on articulation appropriate to the ways we value, decide and act in our daily lives.

Standpoint theory has been criticised from a number of directions and has developed by responding. Yet one area of critique persistently emerges: in 1990 Teresa de Lauretis commented on the inability of Nancy Hartsock's standpoint theory to cope with subjectivity and sexuality, and more broadly with collectivity.[9] The comment echoed other points that found standpoint able to offer perspective but unable to generate collective action across communities. Helen Longino implicitly reiterated the critique in 1993, saying:

> If no single standpoint is priviledged, then either the standpoint theorist must embrace multiple and incompatible knowledge positions or offer some means of transforming or integrating multiple perspectives into one. Both of these moves require either the abandonment or the supplementation of standpoint as an epistemic criterion.[10]

Feminist critiques in the history of rhetoric make it possible to extend the positionality of standpoint, an element that is central to rhetorical stance, not only to methodology and historically appropriate strategies and techniques that a standpoint may have in common with others, but to a detailed exploration of the way different standpoints negotiate social and political agreements and actions that can form conversations and negotiations across and between communities, to considerable benefit.

There are now many commentaries on feminist contributions to standpoint theory. Dorothy Smith in *The Everyday World as Problematic* (1987)[11] cites Helga Jacobson, Meredith Kimball and Annette Kolodny. Sandra Harding, who gives a generous account of extraordinary clarity in *Whose Science? Whose Knowledge?* (1991),[12] cites Smith, Nancy Hartsock and Hilary Rose. More recent accounts, such as the contributions to Kathleen Lennon and Margaret Whitford's *Knowing the Difference* (1994),[13] or Lorraine Code's *Rhetorical Spaces* (1995),[14] cite Harding, along with Donna Haraway and Evelyn Fox-Keller. There is an increasingly interdisciplinary awareness of work drawing on the political theory of Carol Pateman, psychoanalytic object relations theory,[15] on the work on gender-specific training by Carol Gilligan and others,[16] and on the work of Mestiza critics and literary critics from the African diaspora:[17] some of which is gender-related and some race- and ethnicity-related, but all is concerned with the position of those excluded from ruling power.

As I will go on to discuss in the following chapter, there is a profound confusion in the arts and humanities between standpoint and discourse theory, that has led to many problematic albeit sometimes generative attempts to bring together Foucault and feminism.[18] The confusion lies in the notion of what a 'text' is. Dorothy Smith, for example, when differentiating between ruling and non-ruling power, says that non-ruling power is implicated in 'particularised ties' but that ruling power is implicated 'by forms of organisation vested in and mediated by texts and documents, and constituted externally to particular individuals and their personal and familial relationships'. The practice of ruling involves the ongoing representation of the local actualities of our world in the standardised and general forms of knowledge that enter then into the relations of ruling.[19] Here she conflates 'text' with 'representation', using the latter as I have used it but leaving no room in the word 'text' to indicate articulation, or the work on appropriate textuality for those excluded from ruling relations. Hence she fuses text with ideologically allowed or permitted subject positions and raises disturbing questions about the lack of agency in textual work: 'merely' working with words: its continual subjection to relations of ruling power.[20] This is not an unusual conflation. J. Flax in 'Beyond Equality' (1992) carries out a critique of Foucault and Derrida, arguing that a feminist deconstruction of self would look at concrete social relations not just texts, as if 'texts' were not concrete social relations. Considerably more will be said about this assumption over this and the final chapter.

In the feminist critique of the sciences, however, this confusion does not arise and the focus is quite clearly on non-ruling positions. As Alison Jaggar put it in *Feminist Politics and Human Nature*, standpoint feminism argues that feminist theory 'should represent the world from the standpoint of women';[21] that because the standpoint of the oppressed is different to that of the ruling class their knowledge/position is 'epistemologically advantageous' (370), 'more impartial' and 'comprehensive', and hence 'comes closer to representing the interests of society as a whole' (371). Women's contribution will be less biased than 'established bourgeois science' or 'male-leftist' alternatives. These particular claims, appearing fairly briefly if stringently in Jaggar's overview, have been extensively criticised as abstract principles.[22] However, they are given historical flesh in Hilary Rose's *Love, Power and Knowledge*.[23] There she recounts dense descriptions of precisely what 'bourgeois science' and 'male-leftist' alternatives cannot address. The clarity of many feminist critiques is achieved by an acceptance that science, as it always claims, does indeed only deal with representations rather than the articulations of textuality. This has led to a corresponding lack of agency in the feminist critique of science, and to charges by theorists such as Longino and scientists such as Fox-Keller, and Rose herself, that it offers only descriptions, not strategies for change. Those standpoint theorists who do offer a revisioned science, indicate that they understand that it must draw on articulation rather than representation through their consistent movement toward textuality, particularly toward literary textuality: Rose's discussion of science fiction,[24] Harding's world hidden from the 'consciousness of science' that is explored by 'novels, drama, poetry, music, and art',[25] Haraway's use of SF,[26] Fox-Keller's of metaphor and image,[27] and Code's of narrative.[28] Yes, as I shall explore, it is difficult from these accounts to differentiate between first- and second-order representation which take language as a question of adequacy, and articulation which works with language as necessarily limited. Failure to do so has far-reaching effects for the activity of science.

Feminist history of science

Feminist histories of science frequently start with the 'modern' science of the seventeenth century, yet there is little standpoint analysis for this period since it represents an underdeveloped research area, falling oddly between the extensive historical studies of the medieval period to 1500 and the burgeoning field of women in science, technology and medicine from the eighteenth century to the present day. The accounts often duplicate the stories of mainstream history of science which, understandably, attend to the roots of contemporary mainstream science.[29] These accounts are valuable reminders of the strategies science has in common with the liberal social contract: the autonomous individual, neutrally observing nature, achieving objective ends and understanding an absolute truth about the world that can be replicated exactly through rational argument and an experimental method that lets one 'see' the truth of the successful repetition.

Recent work on women and others in sixteenth- and seventeenth-century science indicates points of further research that will be of value to standpoint theorising and which problematise some of the assumptions.[30] For example, the idea that because the labouring classes were not literate in a sophisticated manner if at all, they could not participate in experimental science, ignores the extensive and sophisticated structures of orature.[31] The books of secrets from the medieval period through to the mid-seventeenth century indicate a wide practice of experimental science, and underline the close relation and frequently collapsed distinction between the practice of scientific experiment and of technology during this period.[32] The rhetorical structure of scientific methodology, although more pertinent to the non-experimental science of the medieval and early Renaissance academic institutions up until the seventeenth century, was, however, identical with that of Aristotelian science and with the communication of modern science.

As discussed in previous chapters, the rhetoric of small groups attempting specific end-directed goals and working from previously agreed assumptions, is common to the entire history of science. What is different in the modern period is the resonance of this rhetoric with the new political rhetoric of the liberal social contract – a resonance that both provided intellectual justification for liberal representative democracies at the same time as it turned science into a place where evident political power was analogously constructed as institutional power for science.

The resonance of political and scientific rhetorical strategy takes effect most obviously through the public performance of science. Historians sometimes make the point that science became public for the first time during this period with the institution of, for example, the Royal Society.[33] But science, as the debating field that it had been through medieval institutions, had been just as public.[34] It was the science and technology of experiment, the 'secrets' essential to making a living, that had not been public; and it can be argued that modern science simply added this experimentation to public view. At the same time, other historians have noted the reverse: that science goes from public debate to private practice. Here again the idea of 'private' is complicated. Many scientists publicly performing experiments at the Royal Society, first perfected those experiments in private – in other words at home, often in the kitchens and stillrooms of their households, using the equipment and ingredients they found there, and the knowledge of their domestic companions, familial or servant.[35] Not many of the necessarily amateur because gentleman scientists had the luxury of their own laboratory. It is, however, striking that the 'gentlemanly' class appropriate to modern science forbade performance by women, at precisely the same time that the definition of acting as a low-class activity saw on the Restoration stage English women acting for the first time in 150 years. In both arenas, however, performance is entering into a new relationship with cultural power; one that is beginning to define and limit the individual far more within the structures permissible to citizens and non-citizens of a representative democracy. Men of course practised their experiments in private and only when these were replicable did they go on centre stage to perform them as isolated autonomous individuals of creative genius and power and control.

The sixteenth and seventeenth centuries are a period of enormous fluctuation in the growing understanding of economics and capital and their interrelation with a nation state. Yet even within the discussions leading to the inauguration of the Royal Society in 1662, there is an anxiety about linking modern science with profit. Many of the scientists of the period do explicitly say that scientific knowledge should be public knowledge.[36] Particularly in the field of domestic medicine, which is being taken over by the colleges of physicians, surgeons and apothecaries, there are attempts to publish this 'knowledge' so that families who cannot afford professional help are not without the necessary information/guidance that had traditionally been held by women.[37] However, several of the new scientists were gentlemen but without much property, and either still intent on preserving their 'modern science' as secret so that they could profit from it,[38] or concerned to hide the fact that they made money from it.[39] The early years of the Royal Society saw growing anxiety that middle-class 'gentlemen' members might become too numerous for it to maintain its 'disinterested' status.[40] The later inexorable use of the replicability of modern science by technology and industry to produce dependable and profit-making goods and services, is caught up in this complexity; but it is unhelpful to attribute purposive cause to the rhetorical structure of modern science. What *is* clear is that the participation of women in this newly fashioned public and institutional sphere was considered threatening to its professionalism and seriousness;[41] and that the elaboration of the institution into a breeding ground for ideas that could be exploited by capital was encouraged by wider social, economic and political changes.

The issues thrown forward by feminist critiques of science have surprising correlations with the discussions held by the scientists of the seventeenth century. The receipt books for medical experiments indicate that most practice implicitly recognised that science and medicine involved participant observers and patients:[42] unlike twentieth-century books which usually suggest one remedy, possibly in different strengths, to cure all people of the same illness, sixteenth- and seventeenth-century books like earlier manu-scripts contain many remedies for an illness, and a great many receipts for one remedy. Often clearly indicated is the need to work with the specific situation in a 'located' and what we might now call 'holistic' fashion. Evelyn Fox-Keller's focus on the psychology of the individual to account for differences in science, finds its alterior face in the descriptions of how the four humours affect the individual's preparation of and response to any one remedy. Of course, Fox-Keller is dealing with a discipline that depends for its legitimacy on exact replicability, whereas the experimental work prior to the institutional performance of science is dealing with the impossibility of exact replicability.[43]

Interestingly, the 'secrets' of practitioners are turned by the scientists of the late sixteenth to early seventeenth century, into communally shared and acknowledged receipts – particularly circulated in the manuscript books of women scientists. It has been suggested that Robert Boyle learned this practice of attribution from his sister Katherine prior to advocating it in his *Philaretus to Empiricus* (1655).[44] The activity of science was part of a community of shared knowledge, often explicitly drawn from or in response to social issues. The physical limitations on the dissemination of information at the time tended to result in groups of people working alongside each other, sometimes sharing the same experimental sites[45] and more often sharing a communicative space[46] in letters or visits or conversation between acquaintances. Given the stress by many feminist critiques on the need for working with groups of people, in a 'de-centred' 'non-hierarchical' manner[47] that acknowledges the contextual value of science as a social activity,[48] it is not surprising that the correlations occur. What they indicate is first, the way that the institutional science and technoscience that have developed under the structures of the capitalist nation state have turned this small-group-responsive science into a set of publicly positioned issues subject to ideological representation.

Also of significance is the sharp contrast with the institutional structure and consequent practice of modern science as it has developed between the seventeenth century and today. Apart from the stabilising effects of technology and industry on science, the other primary change that occurred during and following this period was how science was communicated. The increasingly social communication of science brought about radical changes in the rhetorical strategies that scientists used with language to convey the methodology and experience of their experiments, and further, there was a sharp change to how that knowledge was disseminated, distributed and received. Even in terms of manuscript circulation, let alone the print circulation that becomes more expected with the publication of the *Transactions of the Royal Society*, which was edited by H. Oldenburg from the early 1660s, the dissemination of information and experimental knowledge is linked with the class of the scientist. Those people who wrote about their experiments rather than demonstrated or spoke about them, expected a wider audience, and could imagine the benefits because they had travelled. By the middle of the seventeenth century, there was considerable written communication about science, technology and medicine flowing among several distinct groups within the monied classes: aristocratic women, gentlemen, and increasingly aristocratic men. Many people within these groups were committed to the 'common wealth' of knowledge, yet the democratisation of knowledge had to be undertaken through different strategies of dissemination since the small coherent communities were giving way to a much larger constituency. The official medium of dissemination became the representative space of the Royal Society. The people who held that space also constructed a language and rhetorical structure for representing scientific experiment.

Among the many problems of democratisation, as we have seen in politics, is how you deal with larger groups. One response is to represent them, the problem becoming that representation is given to those already visible, already representing themselves in public through for example writing and in performance. In the seventeenth century this was a very small proportion of the privileged population. Hence the 'democratic' representation begins by being a representation of those with intellectual power, usually also with economic privilege, and the structure of the representation will suit their interests, inexorably protect those interests if the structure remains stable and becomes systematised. The process was exacerbated by the developing ethos of ideology which aims precisely to stabilise and systematise in order to protect capital investment and the representations of the nation state that help to define markets. The effects are told differently, but about the same process, in the studies made of the ways in which the automisation and medicalisation of women's bodies became a protected field for male institutional practice.[49]

Modern experimental science became public through writing and performance, and in becoming public at the time that it did it also became subject to allowable positions within the nation state. If technology and industry pinned down the positions for science, it became possible for this to happen through the public communication of science. The books of secrets that document the experimental science and technology of the pre-Renaissance world are often, like dietaries and books of husbandry, lists and lists of lists with little context or explanation. They were specific to a practice of lengthy apprenticeship, to oral delivery and demonstration, and a closely knit community of practitioners. During the seventeenth century, as this tradition was brought into academic science and natural philosophy and as the responsive audience shifted in response to its move from the vocational to the professional, many translations and recuperations of books of secrets attempt to fill in the explanation and the contexts. From Kenelm Digby, who stated that he had had to substantially revise Albertus Magnus' *A Treatise of Adhering to God*[50] because the original writer had concentrated too much on the content and not enough on the form, to the well known concerns of the Royal Society with a proper language for science that have been mentioned in Chapter 2, the new scientists were searching for verbal expression. I suggested earlier that they considered words as bags for experimental experience, and language does appear to have been treated as a second-order medium for science. Even today Fox-Keller asserts that for science, nature is accessible yet beyond language,[51] that there is a continual need to assess the inadequacy of 'a particular language' to science by continually encouraging the 'coexistence' of language with experiment (177).

The rhetorical stance of science

At the centre of the assumption that language will always be inadequate to experimental experience are two related issues of rhetorical stance: first, that any representation that science makes of experience is, as Fox-Keller suggests, merely a tool for intervention (1992: 4); it has no significance in itself; its textuality is second-order; and then, that since what is said about scientific knowledge of the real world is always insufficient, there is no way that appropriate or even adequate context can be provided for it, hence the communicative worlds it constructs are inevitably decontextualised and there is no need to reflect upon them. Despite these issues, the imperative for any scientific communication is to offer an account of an experiment that will make it exactly replicable anywhere, to make empirical experience universally achievable.

Hand-in-hand, like inseparable and desperate, passionate or manic friends, the notion of universal truth in modern science walks together with the rhetorical strategies historically appropriate to its desires. Many people have analysed and discussed the structure of 'objectivity' with its value-free neutrality constructed by the reductive monologic of rationalist argument, but the analysis acquires acute poignancy within an ideological system also attempting universality. If one works within a set of agreed-upon assumptions, it is likely that one's conclusions will share substantial common ground with those of colleagues also working within those assumptions. Indeed, as Scheman points out, 'authorized knowers must have replicable styles'.[52] However, when it becomes difficult, forbidden or impossible to question those grounds, replicable styles become indicators of the things that can be defined as certainly true within the system. The replicability of 's-knows-that-p' statements,[53] characteristic of modern science, is not in itself a reductive occurrence unless one cannot challenge the statement or question grounds of the statement, or make other statements from other grounds to which one will have had to have gained some assent. Yet this is just the problem: replicability within the ideologically stable institution of scientific communication reduces repetition, the source of all communicative negotiation, to cliché, that 'cliquer' of the print type, to stereotype; it reduces an engaged coherence to a model.

Within the stability brought about and demanded by capitalist nation states, the interactions of individuals are constrained by their subject positions. Although these may have been constituted from the acceptance of a categorical imperative, Kant's 'ground of assumptions', they have become, as Sabina Lovibond points out, a totalising structure of 'coercive norms'.[54] More firmly, they become the tautological grounds supporting the tacit aim of those who wield official power to 'preclude action and forestall debate'.[55] The shift is one from consensus to corporately agreed-upon grounds, and from corporate agreement to a lazy corporate self-satisfaction easily eliding into habit and from there into the authoritarianism needed to preserve stability in the face of changing historical need. Nancy Fraser discusses this in terms of a study of Habermas' distinction between communicatively achieved consensus which is explicit, and reflective and normatively secured consensus or habit.[56] Elsewhere she phrases the distinction as one between socially integrated and system-integrated action contexts,[57] which I have earlier termed consensus and corporate rhetorical strategies.

In science, there is a primary mode of communication with the actual world via experiment, yet this is constrained to a significant extent by what can be said about the practices. Given that language is second-order to the experiment, the communication is peculiarly open to the pressures of technoindustry and increasingly to the pressures of firms and the bureaucracy of business, which insist that it maintain the ground of assumptions that determine the stability of scientific development and exploration. The language *is* decontextualised; it builds upon an understanding that there will be a set of tacitly agreed upon grounds, and uses a vast array of rhetorical devices to maintain the fixity of its constructed disciplinary worlds. That this procedure and these devices also form the primary strategies of modern philosophy, especially in its definition of epistemology, indicates not only a broad ideological encouragement for such disciplinary procedures but also points to the second-order treatment of language in that field as well. However, philosophy has no other reality. Whereas no scientist, no matter how they structure their written accounts, expects the actual world to be self-revealing so that there is no interaction between the world and the per-ceiver,[58] the philosopher is not working with an experimental natural science. Critiques of the objectivity of science are critiques of how it legitimises its knowledge and acquires power, and as Sawicki notes[59] this kind of power, being normative rather than liberatory, is then exercised over individuals. Although laboratory life has been studied, it has been studied, I suggested in Chapter 3, for the politics of scientific hierarchy which maintains a club culture emulating the quietist tolerance[60] of gamesmanship. There is very little indeed written on the daily practice of science and its interaction with the real, hence I suspect the fascination with the other tacit knowledge of science: what has not yet been said about its craft skills.[61]

Feminist critiques of science focus on the two forms of tacit knowledge that it obscures: the existence of a set of unexamined, determining assumptions that underwrite its linguistic strategies of objectivity and autonomy, and to a much smaller extent the existence of a field of practices to which it does not give voice. In searching for alternative conceptions of scientific knowledge, the former sometimes begins to give voice to the latter. Often working from critiques of traditional epistemology in philosophy, commentators emphasise the social constitution of knowledge, that for example the notion of neutral, value-free knowledge is 'an ideological fiction'[62] and that the 'values of communities motivate the pursuit of knowledge and shape it'. Or that every society has its 'regime of truth', its interconnection of power and knowledge.[63]

Lorraine Code's analysis of traditional epistemology is based on an ideology of objectivity and dissociation from emotion and value, so that knowledge is neutral and detached, looks to 'natural science' for its models of 'best knowledge'.[64] Yet while her account places the break with such epistemology in philosophy with Marxian and Wittgensteinian theorists, with literary theorists and hermeneuticists, developing interpretative, engaged, participant methodologies, there is no sense of how this addresses the practices of contemporary science. Code continues by precisely outlining the strategies that map capitalist bureaucracy on to the ideology of the nation state, in vocabulary directly resonant of Carol Pateman's critique of liberal democratic politics, discussed in Chapter 1. Those strategies include monologic argument; the privilege of the spectator; the observational mode of evidence gathering; and the knowledge that is produced to facilitate manipulation, prediction and control (122). Code argues that this 'knowledge' ensures the tidy functioning of mass societies (124) but is not 'responsible knowledge'. Responsible knowledge or feminist objectivity is, as Donna Haraway has detailed, situated:[65] it is on-going, located negotiation (111) which resists stereotype and over-swift categorisation (125). It is an empathetic knowledge that treats people, in Code's vocabulary, as 'second-person' not 'third-person' (125–30). Knowledge therefore comes from a 'family of meanings'[66] that in engaging with the world indicates its own partiality. Yet as Haraway again explicates, feminist standpoint is not a weak version of 'perspective'. She says 'Feminist objectivity is about limited location and situated knowledge, not about transcendence and splitting of subject and object'.[67] Furthermore, the standpoints of the subjugated are not innocent but 'least likely to allow denial of the critical and interpretative core of all knowledge. They are knowledgeable of modes of denial through repression, forgetting, and disappearing acts' (255). Situated knowledge does not have a political and methodological imperative to refrain from eclipsing the perspective of others, but a strong sense of position, of *necessarily* partial knowledge.[68]

Where the break with traditional epistemology comes through Marxian and Wittgensteinian theorists, the break with technoscience can possibly be located in an extension of the rhetorical implications of their theories into the strategies and stance of modern science. Sabina Lovibond notes not only the definition by J. Austin in 1970 that 'truth is a value', but also the later Wittgenstein's concept of 'judgment'.[69] A. Tanesini also elaborates on Wittgenstein's development of inferential and justificatory relations between linguistic and non-linguistic acts which 'create' content.[70] Rather than allowing the 'certain' representations of ideology which base their truth on habitual assumptions, the responsive interactions of individuals negotiate the values that are appropriate to their lives, and define the 'truth' of the real world in which they live. This echoes the discussion of judgemental relativism in Harding's *Whose Science? Whose Knowledge?* (153), and of the problem of relativism as an artefact of absolutist epistemology in Code's *Rhetorical Spaces* (185ff). Underwriting the pervasive influence of ideology as the rhetorical ethos of the nation state, Tanesini states that the political structure of society causally influences linguistic practices, and goes on to say that 'epistemology is a game that makes the claims about meaning perform their role' (208). Although potentially relativist, we may choose to reinterpret this Wittgensteinian concept as: knowledge is an activity that sets words into particular actions with which we engage. If so, we are left with a radically empowering notion of textuality that moves us immediately into articulated language, where the necessary partiality of situated knowledge is not a denial of the real or of true and false, or of good and bad knowledge, but an avocation of the necessarily limited boundaries or work of language in communicating our material experience of that reality. Articulated textuality that takes language as limited stresses the immediacy of language to articulating the needs of our daily lives.

When transferring or translating this discussion to science, critics such as Harding, Longino and Rose raise the issue of realism and indicate a deep-seated concern that the partiality of such knowledge will only lead to the plural self-reflexiveness of ahistorical postmodernism, what rhetoric calls the plausible. In terms of a 'critical realism' rather than a 'naive realism',[71] Rose argues for the need to distinguish between good and bad science, hence between true and false knowledge. Yet she goes on to note that absolute truth and falsity, knowledge as 'certain', can only be claimed within a particular community, and that in science, it is only technoscience, the science of technology and industry and business, that behaves as certain (28). Unfortunately, technoscience is the primary representative mode, the main discourse of scientific knowledge, the economic motivator and instigator of exploration and development; hence it is unrealistic not to treat it seriously just because other science has a different ethos. Furthermore, other science is practised so much under the shadow of technoscience, that it is difficult to behave differently from it. The question is asked in earlier discussions by E. Fee and S. Harding, of whether one can have a feminist science without a feminist society,[72] and Rose's analyses adeptly and disturbingly detail these stories. Nevertheless, as she also points out, the positing of a science which can work differently from technoscience is a recognition of the possibility for a feminist re-envisioning of science, Harding's 'successor science'. An analysis of the historical structuring of rhetorical stance can describe the drawbacks of the strategies of modern science as stabilised by the industrial and technological demands of capitalist nation states, and it can also offer a critique that sets out possible alternative strategies for a different stance that has much in common with an analysis of science from the values of standpoint epistemology. Both indicate that other communications and other practices of sciences are possible: both directly address the issues implicit in the two areas of tacit knowledge – the unexamined assumptions and the not-yet-said of craft practice. What rhetoric can offer to standpoint approaches is a way of negotiating the differences and commonalities among several positions; and what standpoint offers to rhetoric is the historical and social contextualising of positions within an enfranchised, representative, liberal democracy.

All the critiques of the objectivity of scientific method, the autonomy of the scientist and the neutrality of the language of science, suggest that if there were a different rhetorical stance to the textuality of scientific communication, this would necessarily address the areas of tacit knowledge that on the one hand allow it to maintain its stability by not questioning its assumptions, and on the other would articulate practice in different ways that would lead to a reconstruction of scientific knowledge and of the organisation of science. The result may be a set of different organisations and differing knowledges, yet these would not be relativist and tolerant of each other but working with part of the material world in ways that may contradict but must be engaged with, negotiating common ground with other partial knowledges in a democratic effort to do good science.

BRIDGEWORK

Articulating the knowledge of those excluded from ruling relations of power, or:
Articulating the knowledge of the enfranchised but still disempowered

Dialogic / dialectics / interpretation: strategies for democratic participation

The concept of knowledge, specifically scientific knowledge, as social raises the issues of valuing, decision-making and action. The local and on-going negotiations of situated knowledge turn a number of critiques toward techniques that tend to be called 'dialogic'[73] or sometimes 'modern dialectics' that can transcend Cartesian dichotomies.[74] Because Marxian dialectics are associated with a transcendence of the dichotomies in the name of a masculine framework, dialogic has more often been used to answer the needs of the otherwise disenfranchised. In a loose sense, many commentaries take the Bahktinian term dialogism as meaning interpretative, engaged and participant. Sandra Harding calls for a hermeneutic knowing that will not distort the actuality being explored by denying the involvement of the knower.[75] Lorraine Code favours the interpretative for its focus on particularity, context and texture.[76] Others follow the spirit of the vocabulary but worry about the connotations of both hermeneutics and interpretation with their own kind of fixity, using instead the Nietzschien/Foucauldian term 'genealogy' that explodes notions of determined gender and subjectivity.[77] Still others comment on genealogy as a process that grounds the practice of knowing in the contingent and historical.[78] But because it is only a process, it cannot offer a critique of ideology.

Engaged interpretation has a different emphasis which all of these uses indicate, although the anxiety over vocabulary does point to the potential for any engagement to become conventional and then blinding. Code (1995) amplifies both the anxiety and the emphasis in her discussion of 'storied epistemology' which depends upon the strategies of narrative to communicate and to train people in 'empathetic knowledge'. Empathy introduces care as a responsive and responsible cooperation, that is conveyed in the structure of intersubjective 'gossip' (151), or narrative that establishes continuities between experience and circumstances, and theories (155). It builds the case-by-case epistemology of analogy (48) and releases the productivity of literary, hermeneutic and interpretative structures. Narrative voice locates theory, knowledge and experience in socio-historical situations and epistemic struggles (156) so that it can 1) determine the criteria for adjudicating knowledge claims, 2) offer methods for analysing society's constructed knowledge so that the marks of the makers and constructions of reality become visible, and 3) be attentive to specific subject matters and open to critical debate (183).

However, storied epistemology is potentially relativist, and Code also notes that temporary closures have to be effected at 'nodal points' to permit action (183).

The comprehensiveness and clarity that Code brings to this field is encouraging. Her critiques give weight to a vocabulary of reason, emotional value and personal experience that is lacking in philosophy, science and literature itself. Yet there remain serious questions to do with the issues of where and how, appropriately, to effect closure, and to do with the difficulty of making the distinction to which she draws our attention, between caring and manipulative empathy, and with the rhetorical stance of narrative itself which is not unproblematically engaged. One of the focus points of interpretative strategies is precisely on the way that narratives usually represent the subjectivities allowed by ideology; however, they offer those in power ways of hiding their own partiality.[79] Rosie Braidotti faces this issue using the work of Luce Irigaray on 'mimesis', to say that we need precisely to work through the sites where subjectivity is constituted, to collectively redefine a subjectivity for women.[80] This work in-between, at Walter Benjamin's 'interface', is dangerous because we need to be so close to the fixed narrative in order to work through it – hence the need, underlined by Nicole Brossard's similar concern with collective action, for communal support, of which I spoke in Chapter 1. Code gestures toward these difficulties most explicitly in her analysis of the structure of stereotype as a self-evident representation that damages the person stereotyped, as well as corrupting the person doing the stereotyping by positioning them in a non-empathetic, disengaged, non-participating 'role' (1995: 101). What is disconcerting in the critiques of knowledge, philosophy and science as second-order textuality is the uncritical gesture to the literary, the interpretative and genealogical, the rhetoric of narrative, gossip and analogy, that in effect ignores Bahktin's own distinction between different kinds of dialogic, the different stances of first-order representation and articulated textuality which I will discuss in some detail slightly later in this chapter.[81]

In her discussion of 'rights' and the welfare state, Nancy Fraser suggests that the excluded sites of non-ruling power generate 'runaway need' rather than determining need,[82] and goes on to say that while this is usually depoliticised by representation it can be politicised by opposition that constitutes new collective agents and/or social movements. Resonant with this vocabulary is Alison Jaggar's concept of 'outlaw emotions' and the 'epistemic potential of emotion': she argues that traditional epistemology offers a structure for certain male emotions, but that other emotions are excluded from the status of knowledge.[83] The underlying movement of standpoint theory is toward a democratising of knowledge, and opening out of epistemic possibilities to all people, and in the context of feminist critiques of science, a radical shift. H. Longino suggests a science shaped by an 'interactive dialogic community' characterised by four criteria which deserve noting at length:

1   publicly recognised forms for the criticism of evidence, methods, assumptions and reasoning;
2   not toleration but change in response to critical discourse;
3   publicly recognised standards for the evolution of theory, hypothesis and observational practice and by which criticism is made relevant to the goals of the inquiring community; and
4   communities need to be characterised by equality of intellectual authority.[84]

These characteristics are identical to Aristotle's corporate rhetoric of science and suitable for small democratic communities, although they leave the problem of deciding when consensus should be questioned because it has elided into corporate stasis. They also leave out the issues of pluralism, particularly of how to engage with larger ideological structures and shift the determining pressure that they exert. In *Whose Science? Whose Knowledge?* Harding brings to bear on science historical analyses of African science and a lesbian standpoint epistemology, in precise and particular reconstructions of possible other sciences. However, she places the 'successor science' within sociology, a suggestion reversed by the detailed work of Hilary Rose, who also takes up issues of race and especially class, to move a re-envisioned science back into scientific practice.

Undemocratic science

Science today is not a democratic process, although it is often perceived as one. Evelyn Fox-Keller asks why modern science is so effective, how can it induce 'the objects of a non-discursive regime to behave as reflections of our own purely discursive regime?'[85] In an answer that forfeits nothing to relativism she concludes:

> The trick that has given modern science so much of its power is the recognition of the need to actively engage this material – above all, to engage it in ways that permit an effective arrangement of its constituents into an instrument that enables the projectile to meet its predesignated target. Perhaps it is the choice of target that we need to examine most of all.
>
> (110)

Because we usually recognise actions in the discursive field of things we can communicate or talk about, we only accept as significant the actions that appear to communicate or behave as if they were within that field. Unlike human beings, the respondent or subject of science does not directly answer back, hence science is in a position of unequal power with respect to the actual world. The natural world is not 'natural', but investigated by people who see what they are doing as part of their actual lives – whether that becomes the study of electricity, of renewable energy, of the genome project. In pursuing science, people are not only engaging with a natural world that cannot answer back, but extending control over it in a way that may and often does affect the engagement of other human beings to it. Hence they are subordinating, and are often completely unaware of, the concerns of those other people to their own. When scientists are drawn from a small, relatively coherent and geographically isolated community, the problem of that subordination is masked; similarly when scientists are drawn from communities all over the world whose education has been sufficiently similar to construct congruent intellectual methods and aims, that subordination is simply not visible. But in an enfranchised society all people have a democratic right to engage with the actual world and to participate in decisions about scientific exploration.

Scientific exploration, just as cultural communication, constitutes the common ground, the epistemic norms, the consensus of our shared values. But as Rose outlines, current science policies in western nation states operate as if democratic enfranchisement had not happened. They ignore the engagements with the natural world that occur in any group interested in issues that lie outside the representations of scientific ideology, only recognising the engagement when it fortuitously enters their discursive field, and often obliterating the source of the knowledge, rendering it invisible if it comes from a group with unacceptable interests and positions. Rose has also documented a variety of ways in which women's contributions to mainstream science have been rendered invisible, initially ideologically but increasingly systematically as consciousness of their contributions begins to upset the balance between the institutions of science, educated privilege and earning power. Harding's analysis of African science speaks of its erasure, erased because it was beyond belief. The ideological representations did not allow for it, so it could not be heard, nor seen nor tasted. Code discusses this dilemma in terms of personal testimony and the 'incredulity' of listeners who have no analogies for it or skills to negotiate with difference.[86] Similar difficulties occur with the slow understanding of the ecological sciences of First Nations people in North America, and more popularly with the slow acceptance of the value of 'alternative' medicine and medical practices.

Returning science to small group practice will not make it more democratic, any more than small Gadamerian communities will return democracy to the state. The fundamental structure of social interaction and responsiveness, in existence in one form or another from Aristotle to the modern period, was profoundly affected by the enlarged democratisation of the liberal social contract which led to the emergence of nation-state ideology, and has been radically altered by the fact of enfranchisement in the twentieth century. These changes have impacted hugely on politics, and possibly because of the collusion of technoscience with capital, also upon science. Yet science does not want a welfare state answer. I suggested earlier that the welfare state could be thought of as the capitalist nation state's response to the suddenly-made-visible representations of women and unpropertied men. Science is now faced with a sharply increasing visibility of women, people of colour, and the less affluent or differently abled, making contributions to scientific exploration. At present the representations or subject positions available to these people are directly analogous to the dependent positions held by welfare recipients in the state. Some analyses stress the way that welfare is constructed to return citizens to the 'master-narrative of temporarily interrupted self-reliance',[87] ignoring that there may be a different narrative in which those people want to participate. Other analyses stress the pathological structure of welfare 'rights', noting that they derive in a tautological manner from the needs of the state to maintain its own bureaucracy,[88] and that they commodify recipients into positions that stabilise politically sensitive issues through administration, law, and therapeutic problems and answers.[89] Still other analyses, in their focus on the welfare state as a way of protracting the liberal democratic contract,[90] go on to argue that the supposed 'special rights' of welfare recipients should be set aside in order to examine the long history of privilege and 'special rights' actually accorded to affluent white Christian men throughout the history of that state.[91] This argument is directly parallel to Fox-Keller's concern with the 'target' of science, with our ability to be controlled by the recognisable representations of scientific discourse and our lack of rhetorical strategies for working on or appreciating those articulations that come from people outwith the relations of ruling power.

Given that Longino's conditions presuppose the achievability of an agreed-upon scientific agenda, that full democracy would be unlikely to reach with the limited resources for scientific work, are we left with the pluralities attested to by Harding's historical analyses which have driven her successor science into sociology? Well, no. The standpoint critique of science is a critique, not merely an analysis. It has offered its arguments about science not from the position of a mainstream history of science, but from first-hand and/or historical accounts of science that not only attest to those exclusions but which articulate the kinds of science that might be practised. Those practices, while not general, are not arbitrary. The difficulty at the moment is that to derive a set of strategies, let alone a policy, for a different kind of science; there needs to be some way of drawing on the experience of those practices without compromising the situatedness of their knowledge. Gesturing to the arts without any clear understanding of the ethical and moral positioning of strategies used by the texts that convey this knowledge, is not going to help. However, as I shall go on to suggest, exploring the situatedness of textuality, which is what rhetorical analysis does, may indeed lead to a better understanding of what encourages a flourishing situated knowledge, and how we might teach, including ourselves, people about it.

Stances for democracy

In a detailed article critiquing the rhetorical structure of contemporary justice, Jane Flax begins by saying that domination – authority and control, the ends of autonomous objective and neutral knowledge – arises from an inability to recognise, appreciate and nurture difference.[92] Her analysis of the failures of postmodernism indicates the questions about communication, persuasion and legitimation that she would like justice to consider:

> 1) how to resolve conflict among competing voices 2) how to ensure that everyone has a chance to speak 3) how to ensure that each voice counts equally 4) how to assess whether equality or participation is necessary in all or in which areas 5) how to effect transition from the present where many voices are excluded, to a more polyvocal....6) how to instil and guarantee a preference for speaking over the use of force and 7) how to compensate for the political consequences of an unequal distribution and control of resources.
>
> (198)

A transposition of Flax on justice and Code on ethics into the vocabulary of rhetorical stance asks us to take knowledge not as fixed or absolute, nor as plausible opinion or relativist, but as good probable ground for actions: reached via debate, discussion and valuation, as one works on words appropriate to the social materiality that will instigate the 'best' response and action. The congruence of standpoint theory with a rhetorical stance directed toward the 'good', offers a vocabulary for thinking about science in a different way, but does science make sense like this? What would it look like to have a science that could 'recognise, appreciate and nurture difference'? I think science can make sense viewed in this radically non-Aristotelian way, eschewing club culture, because its experimental labour and craft parallel the textuality of words from which the vocabulary of articulation derives – although the radical departure required for the democratisation of science may lead us to want to rename this new science. Textuality works not only with the tacit assumptions of the self-evident second-order representative language, and with the tacit not-said-aloud that sometimes finds voice in first-order representative language, but also with the tacit of the excluded that can only be voiced through the work of articulation. Importing a vocabulary or strategies and devices that alert one to a stance for articulating the excluded in science, is not only a communicative change into first-order textuality but a fundamental shift in how science relates its experience of the actual world to society, and would have far-reaching effects on the institutions and ideological pressures which so successfully determine communication within a second-order textuality.

Unlike discourse studies, which can overdetermine individuals into 'subjects' because discourse works in closed systematic worlds analysing the negotiations that occur along the ideology-subject axis, rhetoric comprehends a range of stances. Absolute and determined stances of authority, relativist stances of corporate systematic plausibility, and material stances of probably-the-best grounds of action for all people, are discussed throughout the history of rhetoric as political activity that has communicative substance in words and other textual media (Aristotle refers to music). The notion of an absolute objectivity, a fixed truth, value-free and neutral rationality, has been subject to critique from the start of recorded western rhetoric. As previously mentioned, Aristotle says that it is a game that can be played by scientists and philosophers so long as they remain isolated from society. Plato denied it any value, criticising precisely its social irresponsibility and its deceit: its claim that language can be adequate to reality. The *Phaedrus*, as I have argued above in Chapter 4, criticises the concept of language as adequate because it is either an unhelpful code-making fixity or a justification for plausible inadequacies: the latter is ambivalent because it does at least recognise the first-order textuality of words. But as I shall go on to discuss in the following chapter, this kind of first-order textuality is different from the articulated textuality of Plato's third lover who takes language as necessarily limited and fundamentally material. The distinction is similar to Bahktin's distinction between two kinds of dialogism. The idea of objective knowledge and of language as adequate to reality has been contested back and forth, with stringent critiques from Cicero, Augustine and Agricola among others. As the work of contemporary historians of rhetoric indicates,[93] that contest is always located within the discursive space closest to ruling relations of power, prior to as well as within the ideology of modern nation states. At times in governments, in religion, and in aristocracy, as in science since the eighteenth century, the history of rhetoric documents the movement of this conjunction of power and knowledge, but it also documents the fact that when we see that conjunction it is usually failing. This should alert us to the possibility that today something other than science is moving to the centre of power – possibly information.

Rhetoric is a field concerned with the recognition of and negotiation among differences so that action can be taken: which is why it takes all communication to be a form of persuasion. The three traditional rhetorical divisions of the judicial, the demonstrative and the epideictic, deal with justice or the debate about grounds which leads to their legitimation, with knowledge as 'science' or the debate within grounds, and with the putting into words of grounds. Epideictic rhetoric has frequently been interpreted as congruent with poetics, and when each works toward articulation of the tacit this congruency becomes marked. Plato suggests that rhetoric, which is there precisely to negotiate difference, runs into difficulties when it begins to think of the audience as the 'same': in this many later rhetoricians agree, and it generates their constant worry about absolute knowledge. For Aristotle, the rhetorical strategies of dialectic once more precisely indicate a differentiated audience, while the rhetoric of science means an audience of the same. The basis of classical scientific rhetoric is that either we agree to be the same, or that we agree about the conditions of our sameness that distinguish us from others and then we cope with internal differences within a constrained dialectic, or that we define the conditions of essential human sameness (absolute truth) and deal with the restricted set of differences as a problem – in which case every scientific advance contributes to sameness, not difference. Scientific reductionism is not about finding the absolute, but about treating difference as a problem, as 'wrong', as counter to the 'truth'. In other words it is a stance that uses strategies and techniques that help define the community as more and more the same. Rational analysis does not just limit exploration by the logic of its method, but is precisely a way of impacting on the audience, getting them to present themselves as more the same, asking for acquiescence to the subjectivities determined for them by ideology and a corollary tolerance of others.

If all people are 'equal' in public, that kind of public science requires 'sameness'. This similitude also becomes a source of its power because, within the privileged demands of the liberal social contract, it can claim to be a force for equality by defining universally recognised truths. Hence it is not that there is a compelling force in science itself, but that its ideological intervention through technology is forceful yet not necessarily helpful to an enfranchised population. In the light of the interventions of science, we need an assessment of democratic rights which deals with images of subordinating nature as a 'bad' thing, but which is also concerned with what kind of knowledge about nature is necessary for communal and social 'fairness'. If science were conducted by agreeing to 'recognise, appreciate and nurture difference',[94] it would not proceed so fast, it would need to break down the relationship with technology and industry that business constructs as 'making profit', and to redefine 'successful' for science. This science would need to be far more responsive to group needs: there would be less to get done and more working through issues, except in times of great need which would also have to be defined.

Extending the groundwork laid by Sandra Harding, Hilary Rose has offered a number of specific examples of change in science in 'Goodbye truth, hello trust'.[95] First she outlines recent research which has 'identified' genes for breast cancer, the result being advice being given for 'radical double masectomy as a prophylactic measure where no cancer exists'. In addition one such gene 'is claimed to be particularly present among Jews of Eastern European origin' (10). This layering of science, gender and race is inexorably infusing many areas of the biological sciences, which she delineates, but then moves on to substantiate the claim that 'having feminists in biomedicine makes a difference' (13), with further case examples such as the positive recognition of the activity of women in hunter-gatherer societies. She defines what she calls the 'science wars' between mainstream scientists caught in an objectivist/pluralist dichotomy, which is parallel to the dichotomy of pure truth/relativist truth, delineated by the earlier chapters of this book in the analyses of computing and artificial intelligence by historians and philosophers of science, and parallel to the same dichotomy found in political theorists taking science as a 'best-case' for the epistemology underlying the representative democracies of the West. In contrast to science wars, 'feminists in STS [Science and Technology Studies] show a strong sense of both the difficulty and necessity of building alliances between feminists in and outside the sciences' (20) which has allowed conversation to take place across the dichotomy, and indeed to recast it into a knowledge of 'both/and' rather than 'either/or': a distinction made by many standpoint critics such as Patricia Hill Collins, working on race rather than science,[96] and indeed in Rose's own earlier work.[97] Significantly, Rose traces this activity to

> the normative commitment of late twentieth-century feminism to both diversity and to democracy. There is an evident longing to construct new democratic knowledges of both nature and culture with no excluded 'others' denied their voice. This concept of citizenship without boundaries is a historically new political and cultural project.
>
> (1997a: 21)

While rhetoric has in the past been a place for recognising and negotiating among differences so that material actions as practices can occur, it has also been a place where descriptions about abstract and therefore fixed knowledge give guidance about how to be the same as other people. The great upheavals in the history of rhetoric have arisen from an understanding of standpoint and a revaluation of the material effects and actions of stable knowledge. What is different in kind for the modern world is that the rhetorical strategies and techniques that favoured material practice in earlier periods have been severely disrupted by the introduction of a broader franchise and the ensuing development of the liberal social contract and ideology. Classical rhetoric was not intended to cope with full enfranchisement and the participatory democracy of such large groups of people with so many 'situated' knowledges. However, the history of rhetoric, which is a history fundamentally concerned with texts, is a history of response to an ever-broadening democracy. This is why it is particularly important to look at the rhetorical issues in contemporary textuality, before we can make much sense of the hopeful gesture from science to the arts.

A rhetoric of situated textuality: taking the feminist standpoint critique of the sciences into aesthetics

Science employs rhetorical strategies directly parallel with those used by contemporary representations of power. It works from tacit assumptions about the grounds for its argument, but does so because the actual is more important to it than the representation; it assumes that the actual will always resist misrepresentation. The communication of science is ideological in a technical sense from the start, which is why contemporary political theorists often get excited about it. The stable representations produced by science are maintained by technology, industry and business in attempts at invariant repetition; technoscience becomes a stable source of power constraining individuals into stable subjectivities.

The authority of science comes not just from its perceived effectiveness with the actual, which it had always evidenced even before the modern period, but also from its replication of the strategies of the successful legitimation of political power. Not until the seventeenth century did scientists, by professing, by moving up a class and offering a congruent rhetoric, acquire power. Science, however, acts without the doublethink of politics because its 'subject' is the actual world: the subject can resist in actuality but not in language. The drawback in this for science is that it does not easily realise when it is commodifying itself; only when it comes across different scientific disciplines or cultures does commodification show up. And modern science has always been transnational, replacing Latin as the common European language. Because its communication is so decontextualised, difference is more difficult for science. Furthermore, the multinational global economy changes the ground for science now, just as much as did capitalist nation states in the post-Renaissance period. Science no longer replicates the strategies for the successful legitimation of national power, but replicates the strategies of nation states as they become cultural objects, subject to the multinational economy. It is thereby replicating strategies about whose abuses and manipulations we have developed an understanding. If the nation state in the late twentieth century begins to commodify itself self-consciously, yet institutional science continues to be unselfconscious about its commodification, that science can become a parody of itself with its rhetoric out of control: This is a picture offered in opposing ways toward the cynical and the real respectively, by both Lyotard and Rose.[98]

On the other hand, if the nation state collapses into a recognition only of its social production and institutional science does not, then the relation between science and the multinational economy can continue to mimic that of science under earlier capitalist state power precisely because science claims that it is unchanging, not at the mercy of discursive shift. However, the more science can be cast as social production with a material basis, the less control multinational economics will have over it. Hence standpoint theory which analyses the material basis of social production and historicises discourse, is a profound threat to the entire relationship between knowledge and power, not just one or two immediately pertinent areas. Technoscience is of course rooted in economic power, yet it needs academic science as a breeding ground: institutional science is a concubine to the modern nation state, and there are many children.

A possible way to upset the progression of the relationship could lie in the way that nation states represent themselves. If they encourage an articulative mode for collective social activity or action,[99] then science could be part of that articulation. It may be that the global extent of the multinational constitutes it within its own very large but closed system of claims to neutrality and objectivity. If so, it is vital for the nation state to adopt consciousness-raising rhetorical strategies (discussed further in the following chapter) which will counter the power of 'neutrality' and ensure some notion of 'organised' self and centred identity/entity rather than a permitted representation that reduces collective action to corporate expediency in the name of profit. The impetus for such a critique of the permitted representations of the nation state allowed by multinational neutrality, can come from non-national collectivities across nation states, and science is one obvious place from which critique could be made. Yet it will not be able to do so if it retains only its present institutional structure tied to technology, industry and business, and its corollary structure of public communication that treats language as a second-order system. Other people, particularly the standpoint theorists, have begun to discuss the possibilities of challenging the former. Here I will focus on the latter, and finally I will briefly return to computing because it is within computing science and artificial intelligence that the effects of the lack of first-order textuality in science are most clearly in evidence. However, standpoint theory does make it possible to read science through computing science and AI, as pragmatic or banal *versus* practical and engaged science; and I will follow some of this distinction into the understanding of textuality.

Rhetoric, communication and situated textuality

I have argued that language is usually considered (in)adequate to representation, and as such is either second-order, inadequate code or first-order textuality, language transcending inadequacy. I have also argued that language can be considered limited and that when it is, it necessarily operates in first-order textuality but works on words and *articulates* rather than *represents* reality in what I call, specifically, situated textuality. In what follows, therefore, there are two pairs of different kinds: one element of each is common to the other. Because of the way that they are structured, each element casts the idea of tacit knowledge in a different way. There is the pair based on the post-Cartesian concept of linguistic (in)adequacy to representation, of 1) second-order linguistic code and of 2) first-order linguistic transcendence; and there is the pair that thinks of language as first-order, and therefore believes that it is possible to communicate reality in some way, of 1) linguistic transcendence and of 2) situated textuality. Here, and in the following chapter, I am most concerned to discuss the different rhetorics of the two kinds of first-order textuality, the transcendent and the situated, and to evaluate the implications for action of their rhetorics.

Because science communicates as if language can attempt exact presentation of actuality, it works with a referential system that continually fails. Therefore language is defined as more or less inadequate to the representation, and always second-order to the experiment. Any second-order language is implicitly idealist since it is always apostrophising or gesturing to the possibility of adequacy; and any knowledge dependent on such language for communication will give the impression of *telos*, an end, a fixed goal or an absolute. Knowledge that recognises language as first-order is not denying actuality and the need for reference, but recognising that exact presentation is not possible. This is different from the need for context or for history, except that to take language as second-order denies its own history and context. Language as first-order may border on the relativist if it works only with immediate contexts and neglects the larger social and historical materialities, but language is inescapably relativist if it behaves as second-order communication, because in doing so it denies all context and history. Approaches to language as first-order are open to ideological pressure. One dominant critical perception of language as first-order representation, which is the reverse face of language as second-order, still takes on the notion of inadequacy. Many theories of communication over the last 300 years in western societies, particularly poetic, artistic and literary, have treated art as the place where inadequacy is transcended. More will be said about this first-order concept of language as representation which works best within contexts isolated from larger social realities, but it is important here to distinguish it from what I am calling the textuality of articulation.

Working both with immediate contexts and with history, the textuality of articulation knows that first-order language is limited, that its knowledge is necessarily and helpfully partial. When that knowledge is articulated the textuality becomes situated. Language as (in)adequate is both truth- and certainty-seeking *and* relativist and neutral. Differently, language as limited, or situated textuality, is constituted within history and immediate context, but it also negotiates them. It tells about the actual in probable ways that are concerned with social need. It offers a way to negotiate between the needs of the consensual, or change particular to immediate need, with the stability required by the corporate.

Textualities are congruent with political rhetorics. The dependence that ideology fosters on the need to remember to forget the possibility of agreement and disagreement with grounding assumptions, in effect parallels the work of second-order textuality. Both are locked into notions of language as (in)adequate, and both require the simultaneous existence of a split self – absolute/relativist, authoritarian/arbitrary or anarchic. First-order representative texts that attempt responsiveness within a local environment are similar in their extents to discursive analyses of power: they both understand the need 'to remember to remember' the representations as such, but they do so in systematic and corporate ways that selectively exclude and hence determine many other areas in order to focus on one. This activity of first-order representation is important, because from this relatively empowered stance of privilege the institutions of the state can be criticised effectively; the critique can be heard. But first-order situated textuality that works on articulation and recognises language as limited, is moving toward a consensus politics, that in an enfranchised population needs collectively negotiated positions of difference to ensure a sense of agency.

The amnesia of ideology turns need into inaccessible desire, while the agency of textuality is a continual eroticisation from difference, a continual exhaustion. Work on common ground is often unexpected, and first-order rhetorical strategies are there to remind us about the forgotten as well as to offer certain devices such as *aporia* or *catachresis* that might be appropriate.[100] And this work on the articulation of difference and the excluded is group work, because difference and the perception of it, need community. To note difference in gender, ability, race or class, is to note its representation not its agency; to understand the agency you need to be part of the community, including extended and affected communities. Difference is marked because ideology is not complete, but its relation to ideology is complex. Ideology generates a sense of 'fit': for phenomena, that fit is often called beauty; for the subject, that fit can be called sexuality; and for ruling relations, that fit is governing power.[101] At the same time ideology's child, desire, puts in motion the activity of 'fit'. Fit happens when one can instantiate oneself, commodify oneself into the privileged representative system. The possibility of 'fit' implies notions of adequacy, but also of being outside 'fit', not fitting in, having 'fits', even 'fits and starts'.

The things that don't fit into ideology are not represented and hence are tacit, yet the tacit takes a number of forms. In science there is a comprehensive understanding that the *a priori* are tacit by an agreement reached when one decides to become an institutional scientist. Computing science and AI are obsessed with bringing the tacit into representation, finding words for the understood but not-said assumptions of science. In this it is similar to the sense of 'fit' achieved by science when it succeeds in representing part of the actual world so that it sits neatly in the interstices of ideology, whereas AI attempts to succeed in representing the tacit assumptions of science within ideological boundaries. Neither is similar to the tacit of those excluded from ideology and discourse, the tacit that comes from a difference that needs agency, that realises agency in its work on words toward articulating that need. The tacit, the excluded, the outside of 'fit', may be a preliminary unarticulated essence, it may be the taken for granted, it may be the not-yet-said or it may be the not-yet-negotiated. We cannot easily distinguish among them without an awareness of social practice. In cultural studies of ideology and discourse, representations are often taken *as the case* with no sense of remembered amnesia. When discourse studies engage with ideology and subjectivity they inevitably determine themselves. As I argued in Chapter 1, to be heard by those in positions of ruling power we have to come within the range of representations recognised by ideology, and to do so, we necessarily have to enter a representative system inadequate to our materiality, have to leave things out. Hence our needs become desires, never possible to realise within the system of ruling power. In this overdetermined world, representation becomes simulacrum, all need becomes desire and deprives the individual of agency. Agency is the inability to turn need into desire.

Science and the arts

Parallel with political practice, science needs to find a communication and methodology that remains attentive to how consensus comes about, whether it is still engaged and how to change it when it becomes fixedly coercive. Historical postmodernism asks these questions from the position of the relatively empowered, and stays largely along the relations of ruling power into which its position is deeply bound. Standpoint theory asks these questions from the position of knowledges outside relations of ruling power. And the history of rhetoric asks these questions of both of these positions as well as that of the fully empowered. The commentators within each approach are in effect relatively empowered with respect to the education system of western tertiary institutions, yet only standpoint and some of the recent feminist historians of rhetoric actively encourage recognition of the implications of this position of the commentator, and this recognition leads to different evaluations of science and politics.[102]

If politics and science are working on different strategies for the stance of responsible consensus, so too do the arts and humanities need new approaches. Although as we shall see, the arts have attempted critiques of conventional aesthetics, effectively the critical voices that are heard are institutionally masculine, affluent and white, intellectually tied to discourse and relations of ruling power. Were science to develop its own critical positions, a postmodern methodology for science will simply repeat the adrenaline rush of the chase that gives impetus to so much critique in and of aesthetics, as well as repeat the commodification into fame to which such critique can reductively/seductively lead. Knowledge in all fields has to learn how best to develop a rhetoric for enfranchisement: to recognise that a position is left out, how to articulate it, and how to hear it: how to work on articulating all the embodied but unrepresented and excluded voices.

Institutional science is so entirely dependent on grants and other sources of money, that any critique of it is powerful because it automatically criticises the ideology of the state and the economics of capitalism. The arts do not appear to be based on money – after all, if we can carve an hour out of the day, however difficult that may be, we can all sit down and write or paint. Dissemination of art and engagement with it can be within a local community, but trying to disseminate the values of that community further, so that its needs are represented in society and can engage further, possibly even making an impact on ruling power, does usually involve money. It is costly to enter the institutional systems of the media; when we do so we are recognised as professional rather than amateur, and rewarded with serious attention, able to apply for grants. When we read we rarely think of it as entering an institution for mediation, but the scientist is overtly dependent on the institutional communication of knowledge for their own next step and to affect the actions of others.

People interested in the arts are encouraged by the advertising media to talk incessantly about artistic products and knowledges, but within a system of criticism that tends to universalise from the autonomous subject. Those orchestrated television 'chats' around the table usually bring together people of similar class, gender, education or race, and their job is to provoke and entertain, not to engage. These events have a valuable function, but one that is restricted to critiques of representations that are allowed by the ruling relations of power. However, the events also appear to be *all the story*. When a book surprisingly and suddenly sells many copies, is this populist or popular engagement? We have no recognised way of telling the difference. Indeed it is even difficult to do so within our own readings.

It may be that modern science has been unable to accept the concept of first-order textuality because it has been tied historically to a poetics and aesthetics based on concepts of original genius and isolated independence. We tend to forget that as modern science was forming itself, so was modern aesthetics in terms of the 'humanities' which were and are deeply imbricated in the liberal social contract, with its notions of the abstract individual exercising power over the word as well as the world. This aesthetics is not always or only about engaged textuality but also about the self-display involved in that textuality. It is concerned with authority and progressivism based on that authority, for example Pope 'improving' Shakespeare's words.[103] This activity is about formal rather than explicitly social engagement, and the formal is by definition learned social engagement within convention. Despite the fact that these issues could be a description of some science, there are salient differences. The scientific genius says he is discovering a truth about the world, a truth that anyone may replicate, whereas the critic says that art can only happen once. What makes the latter possible is the first-order textuality of the arts that ties its communication to an individual, foregrounds the human intervention into the real. The different approaches to textuality result in quite different attitudes to ownership over the word. The hierarchy of communication of knowledge in the sciences is patrolled at one extreme by patents, and at another by the communality of the laboratory which attributes a scientific paper to many individuals, thus lessening the 'ownership' of any one idea embodied in experiment, despite the hierarchy of the authorship attribution in printed papers. The dominant literary arts in conventional Euro-American culture, because they are first-order texts in a printed medium based since the eighteenth century on valuing individual authors, are patrolled by a copyright system based on the order of words. It is as if the scientist and artist enact the ambivalence of the liberal individual: universal and hence infinitely replicable at the same time as autonomous and unique.

Science also appears to think that it makes claims on reality that are different to the 'fictions' of the isolated individual. But this misunderstands the varied actions of fiction and aesthetics, which more generally may be to 'copy', yet more often in the post-Cartesian world is to transcend the gap between actuality and words, and occasionally is to engage with the material reality. Post-Renaissance aesthetics stresses that textuality 'gets you closer' to reality by giving you more power over representations of reality; it is more adequate and in moments of 'beauty' achieves adequacy to the real – verbal activity that is usually alien to the communication of science although entirely congruent with the way autobiographies and diaries of science recount successful experiment. But textuality can also articulate recognitions, relations and negotiations; it can invite engagement with the real from its audience, and embodies a reality in itself: in the way it articulates reality it is about the communications and constitutions or constructions of society, ruling and non-ruling. This kind of textuality derives its power from the simultaneity of its persistence and instability that brings about a discursive disequilibrium at the same time as it articulates need. It is the power of the group or community working on words, not the power of a subject over words. Such textuality does not take objectivity as absolute power over words, but as value learned from negotiating the resistance of the real.

The deep distrust in science of the textuality understood by the arts and humanities may come about partly because of the way the artist is displayed by conventional aesthetics as the unique producer of knowledge, while scientific knowledge depends upon repeatability by many people. But arts people are trained in a different kind of textuality, and with respect to accepted knowledge they have a different and excluded standpoint. It is necessary to say, in the face of the power of scientific explanation, that there are other ways of knowing that may be *better*. However, this cannot be said without a moral and epistemological basis for an aesthetics different to the traditional post-Cartesian framework. The public understanding of science initiative needs a complementary initiative in the public understanding of the arts. Just as standpoint theorists who look at science have developed an articulation for a feminist standpoint, so standpoint theorists who look at the arts need to come up with a feminist standpoint. Yet there has been great reluctance to do so. The arts has for example a canon, and a large field of diverse literary criticism and theory addressing textuality that is largely white, affluent and male. Both need to be opened out toward standpoint's understanding of shared social space, and to extend the history of rhetoric into making shared common ground in the arts. This will be discussed in more detail in the final chapter.

Doing science differently

What would it mean to do science textually differently, to take a feminist standpoint approach to the rhetorical practice of science? Women, because they have been and are in a position of exclusion or marginality with respect to institutional science, can make different textualities. First, the strategies for articulating grounds from a position of exclusion are different from those used by people in positions of inclusion, although both may articulate realities that ideology renders invisible. Second, there will be different understandings of resistance to ideological representations, of the extent or degree of exclusion, of appropriateness. For example, a common ground articulated by someone openly included within ideological representations, is likely to address an element excluded from representation in the lives of many people with privileged status. The articulation will probably gain widespread agreement from those people, however relatively empowered they may be, and it may elide rapidly into the habitual and taken-for-granted. For example there is a substantial difference between the social response to the middle-class female university graduate arguing for the right to abortion and the response to the working-class unemployed woman arguing for the same right.[104] This is the inverse of the rhetorical force of Rose's claim that women doing biological science have made a difference: without the women there the' activities and conversations that open out excluded knowledge simply will not occur.

In addition, agreement to common grounds is the basis for epistemology, but there are different strategies for agreeing to common ground, and some strategies may leave grounds open to reassessment. Furthermore, there are also different ways of constructing a common ground that may or may not leave it open to change. The structure of the common ground and the method for its legitimation will have an impact on the effects of the knowledge, yet it is impossible to separate them. Structures in themselves are not inherently fixed, discursive or dialectical/dialogical, but can only acquire significance in a social and historical placing.

The textuality of institutional science 'succeeds' when it draws on and simultaneously hides and forgets the discourse authority that mobilises it. Because standpoint feminists recognise different sets of prior authorities for women, from non-ruling relations of power as well as from ideology and its nodes of discourse, successful covering up of the latter (ideology) still leaves the former (non-ruling relations). A new science could try to cover up the non-ruling relations – such a strategy would have effects on practice although it would 'look like' normal science. Perhaps a new science should explicitly value the open communication of non-ruling relations, in order to contextualise its conventional success. Perhaps the new science could go further and open up institutional convention to reassessment within the broad social structures that impinge on the lives of the excluded. Each rhetorical position begins from the assumption in feminist standpoint theory that women have different sets of prior agreements and hence bring about a different kind of knowledge.

In the first instance these differences can simply make it possible for women to be aware of the grounds of science in a more foregrounded manner that encourages revaluation. The difference can also lead to women being in a position to bring differently authorised grounds to scientific knowledge and textuality, after which the grounds may be assimilated and covered over or may come from areas of non-ruling power that fundamentally imbalance the whole ideological/representative procedure for hiding assumptions. People from excluded groups can bring about strong objectivity: an objectivity that is not content to sit within conventional and pathological textual structures, but an objectivity that is strong precisely because it can work from non-conventional as well as conventional grounds and still have an effect, possibly more comprehensive, on the actual. Feminist standpoint approaches can also, more radically, ask science to engage in the textual production of common ground, the articulation of value that is the responsibility of all people to their community and society. This step necessitates a first-order textuality of articulation because only that textuality invites the audience into a responsive engagement – what the standpoint theorists stress when they talk about interpretation, hermeneutics and participation in the communication. But what does it mean to be a feminist working in science and aware of textuality? To begin to answer this I would also like to ask: what does it mean to be a feminist working in computing science and AI and aware of textuality as first-order?

Science and computing science

My own work as a scientist, rather than my education, was as a technician. The lab technician carries out, does, the experiment, and may occasionally suggest but will rarely devise their own experiment. I learned by apprenticeship to a succession of people in what can only be called a patriarchal domestic structure, with attendant systems of harassment and protection. Some people would differentiate between a scientist and a technician by saying that science is precisely not technical but hypothetical and virtual, but science is engagement with the phenomenal world, the actual world. The scientist may work in the hypothetical, but without contact with the actual world, without impact or effect on the actual, science would lose its adrenaline rush. However, the adrenaline is one area of differentiation. As a technician I frequently did not know what results were needed to prove an hypothesis. Doing the experiment gave me an entirely different sense of pleasure, small things like achieving a precise measurement with a balloon-pipette with just one squeeze of the hand, or filling an immunoassay dish with one drop per hole in a neat swift pass, or indeed, repeating, exactly, the same experiment for the twenty-third time. But my associated scientist would be transported into the stillness of certainty when hypothesis found representation in experiment. This disjunction has always underlined for me the accuracy of the saying that we only see what we are looking for; that scientists realise in experiment only that which can be recognised.

Of course there are many other differences between the scientist and the technician. A scientist is institutional, surrounded by a club culture with its own hierarchy of grants, applications and rewards; also, and importantly, it is the scientist who communicates to the public. The technician has their own institutional space and hierarchy, but the rewards are less prominent. And most of all, the technician is not responsible for public communication: when the technician achieves this position they may even be accorded, temporarily, the status of scientist. Even the technologist, engineer or applied scientist has less need to speak to a public audience, being answerable largely to industry and finance. The primary drive of institutional science is to understand and recognise 'fit', to define the beauty that transcends the inadequacy of language to the actual. It is a kind of mastery over the actual that implies control if it is applied, but in itself is held as truth. Here the opacity and resistance of the actual world is reminiscent of Chaucer's *Wife of Bath's Tale*, in which the woman's resistance to commodification into 'beauty' is a sign of her own 'mastery' over her individuality.

However, it is more difficult to make the distinction between scientist, technician and technology in computing science. When talking to computing scientists I have often heard it said that the moment AI manages to represent an assumption from science, computing scientists immediately apply it. The primary drive of computing is to reach an end that has effect on the actual world; in this sense all computing science would be engineering. Yet while computing tries to behave like a science it has no actuality, no 'world' except for its medium. When I learned about computers, and re-learned as different systems came along between 1968 and 1992, I found myself learning grammatically correct 'languages' that were not as interesting or flexible as mathematics. My experience was frequently parallel to Turkle and Papert's descriptions of people who feel as though they are being asked to be 'someone else' yet they could not quite see the point. Many practitioners of computing learn that language is first-order in the sense that without that language there is no science. As someone interested in textuality I should have been happy with this, but computing science behaved then, and still largely behaves, as if the way science uses language is related to what science is doing, as if the strategies of its second-order language are significant for the effect on the actual. Computing works with media as if they are self-evident, it works with second-order textuality as if it were in negotiation with the actual.

When we use computers we do not contact actuality. In the sense of the hardware and software being an artefact of other real human beings, yes, we do. But the transparency, which with language is taught through grammar, logic and poetics, informally from childhood and formally for ten to fifteen years, the transparency needed to recognise these rhetorical structures is simply not there in computing for the humanities for any but the most knowledgeable user. This isolation is compounded by the sense that the language we are using is code rather than text; certainly there is little understanding of code as first-order textuality. There have been a number of commentaries on gender and computing that focus precisely on this isolation, particularly by Sherry Turkle and in her work with S. Papert, that draw on Carol Gilligan's research on the way boys and girls receive training in different modes of argument and behaviour, so that their attitude to moral responsibility is also different. Turkle and Papert suggest that girls prefer relating to the computer not through abstract reasoning but by way of concrete 'bricolage': 'bricoleurs have goals, but set out to realise them in the spirit of a collaborative venture with the machine';[105] bricoleurs have 'conversations' and 'negotiate' with the machine.

Turkle and Papert make the reasonable assertion that if computer textuality could become more 'concrete' and less abstract and isolated, then the computer as an expressive medium 'encouraging distinctive and varied styles of use' (157) could help science more generally to open out to textuality. But they also imply that visual icons are more concrete than verbal text; and that object oriented programming, in its ability to offer context, will make computing textuality more engaged. They underwrite the claim that mathematical, formal and linear structures are masculine as well as more abstract and less engaged. I confess I have personal problems with this. As someone who is trained in verbal and not visual textuality, I find icons less helpful to work with. Object-oriented programming is an interesting step but not one that will necessarily engage the text. Its rhetorical strategies mimic the selective contexts of discursive analysis. And finally, as someone thoroughly trained in mathematical, formal and linear structures, I do recognise that in themselves they can be concrete first-order textual devices – an aspect hinted at but not expanded upon by Turkle and Papert when they mention that 'adult' mathematical reasoning is very concrete (130). In other words I want to resist being told that I am not behaving like a woman because I have been trained to do these things well.

The possibility that education may give women the facility for 'scientific' thought, or men for 'bricolage', exacerbates the need for an historical understanding of the political pressures put upon science and computing science. Gendered issues do arise from the congruency of the rhetoric of science with that of an ideology focused on and for, a small percentage of white, affluent, usually Christian men. Science has indeed used a formal, linear logic to build self-contained explanations of the actual that appear to be powerfully connected to the world through the effectiveness by which they contribute to technology, industry and business. And certainly, computing science deals with this historical use as if it were essentialist, and often uses the strategies in an isolated, reductive and controlling manner. And women's fear of the computer, which may derive from this background, is extraordinary. This fear ranges from that of of entering a 100-seat computing cluster and being one of the only women in the room as you try to complete assignments late at night, to fears that it is so 'powerful' that it may emit damaging energy. Given the unexplained clusters of miscarriages among women data-processors, the fear is immediate. One of the very first devices I had to use to overcome this latter fear was to get a security officer from my then university to publicly perform a test on the standard machines for a variety of radiations.

Yet we know that in the 1970s more women used computers than men, more girls took computer training than men. Only when the computer began to be perceived as a powerful technology rather than an extension of secretarial work, which is often treated as second-class work, did boys begin to overtake girls in training and then in use.[106] More difficult, during the several years in which I taught humanities computing in the 1980s and early 1990s, I was struck by the way the cultural power of this machine intimidated both male and female users from the beginning, so much so that although used to a wide range of other machines, many people believed that a computer would turn on by itself. I was also appalled by the evident hatred, a strong emotion, which computers evoked from both men and women when a task was not completed as expected. When this hate became too evident, I conducted my teaching by subjecting the computer to a constant stream of ridicule, which the students usually self-consciously recognised as a rhetorical device that embodied and enacted but also contained their fear, yet which was not, however, going to be effective in the long term.

Teaching is often an acute guide to how a discipline represents itself to its public – hence the analysis of computing science textbooks in Chapter 2. Although Donna Haraway notes that scientists 'tell parables about objectivity and scientific method to students in the first years of their initiation, but no practitioner of the high scientific arts would be caught dead acting on the textbook versions',[107] there is a world of scientific labour between the student and the small group of high scientists, which does indeed conduct itself as if these parables *were the case*. And there is no doubt that the arts and the sciences have fundamentally different attitudes not only to what is taught but also to how it is taught. This may partly explain why, just as most of the critiques of science from traditional philosophy, ahistorical and historical postmodernism, and feminist standpoint, gesture toward the arts as a site whose rhetorical strategies may halt the reductive progress of technoscience, just so many of the feminist critiques of computing make similar gestures. Several papers to come from the Women into Computing group cite the less aggressive challenges of humanities pedagogy, the group-work environments, the 'soft' programming skills of arts users and women.[108] Yet again, in the commentaries on computer use that draw on the arts, I am worried about the conjunction of 'the arts' and 'women', as if men in the arts are somehow less male. What is ironic is that most of the studies are done on arts students learning about computing developed for sciences rather than humanities. It is unfortunate that the humanities' response to computing has tried to make the arts more scientific in order to access the power and money beyond computing science, rather than try to open up computing to a focus on text, context and history.

The concept of hypertext nests that I developed with Margaret Beetham, involved a self-conscious attempt to enable the student to understand how the structuring of knowledge about periodicals and about hypertext itself, comes about. Central to its pedagogical use is its ability to encourage students to discuss the ways in which they are producing knowledge. If the argument I made at the end of the preceding chapter placed the hypertext methodology within a rhetorical framework, it is possible also to place it within the framework of feminist standpoint theory. To do so would be to underline the importance for the students to position themselves when they begin the discussion about the production of knowledge, to acknowledge other people's position, to develop strategies for agreeing to common grounds that will enable the knowledge to be structured on recognised values. The students do not, indeed are actively discouraged from any attempt to construct a fiction about having the same intentions or position toward the knowledge being produced, and similarly the result of their structuring linkage does not aim to be unitary.

In this, the makers of the hypertext nest are trying to build commonalities that are specific to each project – for example, they may choose to build a nest that constructs an environment from the periodical for the study of women's rights, or, differently, an environment that emphasises the social contexts of production. They are engaging in consensual discussion and self-reflexive activity, yet arresting the process in the production of knowledge, here a particular nest linkage, which has specific agreed-upon reasons for being made. The problem, as I noted earlier, is how to prevent that arrest of the consensual from devolving into corporatism or reified authority. And this is where the positionality of standpoint is so important, because that positionality means that the knowledge is always probable, not absolute.

Like deconstruction, which Derrida conceived as a fundamentally moral project that would enable individuals to try to dismantle their subjectivity so that they could position themselves for discussion and action on broader grounds than only state representation and cultural determinism, consensual argument was intended to move people toward shaping common values so that they could do things, effect action. Defensively, Gayatri Spivak points out that deconstruction is not against 'unitary action' but against it being monumentalised[109] – in the words of Kristeva's essay on 'Women's time', where monumental time lurks as a warning both to linear and to cyclic time. Just so, the knowledge generated by computers, whether it be rational or bricolage, is not necessarily reified if we work on structures and strategies (structures as strategies) to encourage articulation of individual position, which necessarily deals with the differences of others.

This work is particularly important for computing at the moment, and not only because it is such a new medium. If science replaced Latin as the international culture of nation states in Euro-America during the modern period, so computers have in many ways superceded science in their ability to form a transnational culture for globalisation. Computing has to resist the pressures of business and industry and to keep its strategies probable, or there will be reifications everywhere again. If it is structured on the evaluation of immediate practices, computing becomes an active medium by way of which human beings communicate to other human beings. The next chapter will go on to suggest that the intercommentaries which arise when we allow the strategies of the arts, the sciences and computing to filter the light on each other, indicate that a radical critique of western aesthetics is long overdue.

6 A feminist critique of the rhetorical stance of contemporary aesthetics

Alternative standpoints

Standpoint theory is an area of discussion I became caught up in mainly because of its profound challenge to the sciences, including computing science, that I had practised for many years. It made sense to me partly because of the congruency with issues such as objectivity, isolation and the fact/value separation with which I was already concerned,[1] and partly because it so clearly fitted into a gap in contemporary rhetorical theory – a gap delineated by comparison with the consensus-making stance for participatory democracy within classical rhetorical theory and its subsequent history. However, to extend the emphasis of the earlier point, I began to realise a little late in the day that the changes in political practices that had arisen out of enfranchisement had impacted not only on science but also on the arts. Now, my on-going exploration comes from attempts to find ways of talking about written texts that have not received conventional or traditional aesthetic commentaries, which are not perceived as 'literature'. This exploration is the self-consciously political action of an educationist: we can all enjoy and work with texts, but without a sustained attempt to find a vocabulary or strategy for saying or communicating or writing what it is we value in the texts and why we spend so much time working on them, they slip away. The written texts do not get reprinted, nor even in some cases printed; they do not get circulated and read by others. Therefore what we value becomes thin or confined, does not extend out and inform other ways of life.[2] The attempts here are political in the sense of legitimacy: finding ways to recognise and agree to accord value; and in the sense of agency: finding ways to repeat that value in actions.

SITUATED KNOWLEDGES: SITUATED TEXTUALITIES

The feminist critiques of objectivity in the sciences are based on three elements thrown into sharp relief by computing science and artificial intelligence. First, the autonomous isolated individual; second, rationalist analytics in the service of absolute value-free logic; and third, neutral objectivity with verisimilitude and exact replicability primary among its several strategies. The critiques elaborate on the ways in which these elements erase all people who do not fall within the boundaries of the representational parameters – including women. Yet, to recapitulate, isolation, absolute neutrality and objectivity have been the hallmarks of science since the classical period. Whenever they have strayed into politics they have consistently been subject to critique by rhetoricians. Only when science and politics form an alliance in the seventeenth century does science begin to make claims on the universality of the modern liberal man. The movement of science from debate into performance effected by taking over technology, happened just when nation-state capitalism was beginning to understand the potential in technology and industry for profit. The rhetorical strategies of the communication of science, being so congruent with those of the social contract, promised the liberal state the same success with the real that science achieved through experiment. As the preceding chapters have outlined, the congruency has prompted many defenders of neo-liberalism to turn to science as a 'good' example of how the social contract can efficiently work.

Standpoint theory takes a different position. It asks: could you do science differently if you did not erase those people outside the institutional representation? And if so, how could you do it differently? Drawing on discussions of Marxist theory and frequently referring to the distinction between ruling and non-ruling power made by Dorothy Smith and others, these questions are specifically bound to the claim that women doing science do it from a special position outwith the systematic; hence they can be more engaged and engaged with different things. Women in science can more sharply expose its assumptions because they are not part of the prior agreements, hence they can be more objective. And precisely because they are not part of the prior agreements about the laws of science, they may prioritise other agreements, including agreements about strategies for dealing with the actual world that are different from those used by institutional science; strategies with a different understanding about power relations between the observer and observed and generating different kinds of knowledge. Here I would want to emphasise again that these different kinds of knowledge are not simply those relationships of intense beauty that have been documented by a number of practising scientists. Those private relationships are satisfying but they are premised by the positions of privilege within which most of those practitioners work. Other kinds of knowledge, possibly not so completely defined and defining, can come from people who work in a wider negotiation with other people, collectively working with science as a text, as a medium for human communication.

Many developments of feminist standpoint theory end by focusing on the absent textuality of science, its inability to communicate through an engaged and interactive stance, either with its topic of study in the actual world or with its audience – although as the previous chapter tried to illustrate, the inability to engage its public audience has inflected both that audience's understanding of science's relations with the actual, as well as the understanding of many scientists involved in the technoscience at the service of and subject to industry and business. Consistently, the critiques of science turn to the arts, on the basis that the arts are there precisely to offer strategies for interactive and public communication. The arts have traditionally refined textual approaches to offer contexts and supply the situatedness for situated knowledge. Rose, Fox-Keller and Harding, as noted in Chapter 5, all turn to genre, metaphor and narrative; and Code, working explicitly from these feminist standpoint critiques of science, elaborates on the need for narrative to do what cannot be done from the subjectivity of isolated man, value-free logic and objectivity, and develops the use of story to educate one in the sensitivity to collaborative and ecological connections that builds epistemic responsibility.[3]

PROBLEMS WITH THE GESTURE TO THE ARTS

Yet in the gesture to the arts there is an implicit suggestion that narrative or metaphor or fictional genre are necessarily 'good'. Rita Felski articulates the problem with this quite clearly in *Beyond Feminist Aesthetics* (1989), describing in an unselfconsciously rhetorical turn that no technique, strategy or genre is in itself a 'good thing'.[4] Yet in that understanding and with an acute sense of the growing importance of autobiography studies, Felski moves aesthetic value to the recipient. In a contemporaneous move, Janet Wolff moves the aesthetic focus to the institution.[5] As Felski was foretelling, the notion of the arts' strategies as 'good' leads to a philosophical hiatus, a gesture toward the arts with no concept of the situated textuality in the arts that is directly analogous to the concept of situated knowledge. I am using rhetoric, as situated communication, to make a bridge between the two. The lack of any bridge means that an element central to contemporary western aesthetics, the notion of language as inadequate to representation, which underwrites the absolute/relativist divide in the criticism of the arts, goes without critique. Certainly the critique of aesthetics from a standpoint or situated knowledge position is markedly thin, perhaps because the textuality is gestured toward and not pursued. Where textuality is pursued in detail by critics who use the term feminist or standpoint of their work, is in studies of autobiography and autography.[6] For reasons of long-term moral and ethical change in the way we value literary work, which I personally recognise as more positive than the direct critique of a dominant ideology with which I am engaged here and which compromises my work in that ideology,[7] the studies in auto(bio)graphy deal implicitly but not explicitly with contemporary aesthetics. I have a different shorter-term political aim, and there is a need for both.

As this chapter will go on to indicate, the explicit grounds of auto(bio)graphy studies turn toward rhetorical stances for negotiating the relationship between the self and community. Jeanne Perreault argues that 'contemporary feminist autographers...make "I" and "we" signify both continuity with an on-going life in a body and a community, and a dissociation within that life',[8] so that there is an 'interrelation of self and community' (7). One example that she cites is the way that 'personal speaking shifts to communal speech' in the work of Adrienne Rich (43). Another example might be Patricia Hill Collins who, in *Black Feminist Thought*, distinguishes among levels of oppression in personal biography, the communities or groups of race, class and gender, and the systemic level of social institutions.[9] Standpoint theory in the sciences, in sociology and social policy, in politics and in philosophy, is doubly concerned with that negotiation between the personal and the community, and with its interaction with dominant systems of power. My own work here is an attempt not only to contribute to the bridgework, begun by Lorraine Code, between the textuality that is articulated by auto(bio)graphy studies and the need to understand that textuality in other disciplines, but also to open up the implications of such positioned textuality for work in institutional areas that are based on the dominant aesthetic approaches. In other words I am overtly bringing an outsider's standpoint to give a particular agency to an insider's actions. At the same time, as someone who has learned from the history of rhetoric that not all standpoints are morally acceptable, I am concerned to emphasise that textuality is not always conducive to moral or indeed ethical awareness. Much negotiation between the self and community is authoritative, it imposes rather than articulates, and so does a considerable amount of textuality.

To give an example from just one area that I know well, fantasy and the fantastic: studies of the rhetoric of this genre consistently analyse it as being based on a rationalistically persuasive narrative structure that always at one point asks the reader to remember to forget the real, to accept the isolated fantasy world as universal, bolstering the acceptance by techniques emphasising visual verisimilitude and consistency in realism.[10] Because of the historical materiality of these strategies within a framework of western education and communication that accords them a privileged status, the stance of fantasy is always to effect autonomous control over 'natural' and social worlds, and is congruent with the conservatism of satire. It is evident that the elements of fantasy as a genre are parallel with those criticised in the 'objectivity' of science. Although some fantasies depend upon the moment of remembering the unreality of the isolated world, and some that have been called 'fantastic' depend upon maintaining the uncertainty of this remembering,[11] many isolated worlds are constructed for uncritical acceptance, and fantasy is a notoriously escapist genre. Despite the ability that the genre evidences for building alternative, possibly utopian social practices which can inspire and give hope to people, it does so in an ultimately paternalist manner.[12] The stance does not encourage the reader to build their own alternative and leaves them dependent on strategies that seek control over and are subject to, rather than find engagement with, the material conditions around them.

It is possible to read against the privileged status given to fantasy strategies, to deconstruct their control and even to criticise their manipulative pressure. We can learn how to do this with sufficient formal and/or informal education, but we need some kind of social positioning for the work. If we read against a pornographic fantasy and recover it as an analysis of obscenity, without social location it can simply be a self-indulgent exercise that may indeed be harmful, may brutalise other readers and render them insensitive, if passed off as a generally accepted reading while in effect lacking any contact with communal debate.[13] A more pertinent example could be the question of whether or not to recover Margaret Atwood's *The Handmaid's Tale* as a text that works for rather than against feminist debates and needs. In my experience the attempts to read this text against its potential for controlling fantasy is productive and helpful. The issues raised are taken up by colleagues, friends and students in many other discussions. The text also invites such a reading, providing clues and teaching the reader subversive cross-genre strategies that are directly applicable to the structure of the discursive sites in which we often find ourselves. As frequently happens with fantasy genres, the further in chronological time the reader moves from the writing of the text the more mechanical the control over the other world they attempt, and the more we need these engaged readings to enjoy the text. Furthermore, they are not self-evidently attained as the history, and only recent critical recuperation of Mary Shelley's *Frankenstein,* attests. Readings that recover engagement with texts have to be laboured over, usually in communities of friends, colleagues and the wider audiences made possible by the process of publication. Much more could be said, but the intention of this example is to emphasise the extraordinary subtlety and complexity of textuality which can include, often unawares, exactly that pretence to universal autonomy that texts in the arts are often uncritically held to counteract.

If the artistic texts are also able to carry and impose precisely the same elements as reductive science, why has there not been a parallel and through-going critique of aesthetics and criticism as they are taught and/or disseminated in the same institutional structures of the western world? With some editing we could paraphrase the opening of *The Radicalisation of Science* (1976) in vocabulary pertinent to the arts and ask,

> what features characterise the present social function of art in capitalist societies? And argue that, today, art has two major functions, as part of the systems of production and of social control. Art has itself become industrialised through the mass media and enmeshed in the machinery of the state. We can go on to examine two myths, the liberal academic

myth of the autonomy of art and criticism, and the orthodox belief in the inevitable contradiction between art and capitalism, and show that neither accounts for the actual development of art and criticism and arts policy – the management of art. The question is whether art under capitalism represents an unavoidable and fatal attempt at suppression and oppression either individually or via the mass media, or whether it can be confronted as a paper tiger to make way for a genuine art for the people.

Is art part of an engaged process or purely a means of reproducing allowed representations? If the latter then artists and critics, whatever the contradictions within their role, cannot be regarded as working on representing the excluded, but primarily as within or associated with the relations of ruling power, either by assisting in the structural mainte-nance of capitalism, like lawyers or accountants, or as transmitters of its ideological values, like teachers or journalists. That is, they will in gen-eral find that the demand to remember to forget, that forges the contra-dictions of capitalist society, does not oppress them but serves to protect their privileges and position. On the other hand, if art is part of the productive process, 'artists' and 'critics' are really relatively disempow-ered with respect to ruling relations and they sell their work within the capitalist system in parallel with other works; like other workers/people they continually remember the conditions of their disempowerment and can become alienated from their products.[14]

And, as Hilary and Steven Rose concluded then for science, I would argue that the 'arts', or art and criticism does both. The significant theoretical difference between this New Left Marxist argument and the debate in science today is the standpoint elaboration of the power of those excluded from or marginal to relations of ruling power. Those disempowered or excluded from ruling power may become 'alienated' from the products expected of them, but standpoint has developed a vocabulary for articulat-ing the agency of those affected by but disempowered with respect to relations of ruling power. People may also work on those 'products' needed by the collectivities in which they live day by day. Such collectivities have relations of power, but these are negotiated, agreed to and acted upon, not imposed by state institutional systems.[15] Another material difference is in the exponential extension and visibility of global capitalism which, as has been argued throughout, inflects the relations between the individual and the state institution in different and as yet little-analysed ways. However, what is striking is the immediacy and pertinence of the paraphrased questions for aesthetics and criticism today, particularly: what is democratic art? What is art for the majority of the recently enfranchised population that has for the past century at least, made claims not only on political power but also on artistic power and aesthetic value, and has received, if at all, only grudging legitimation?

The larger part of secondary and tertiary Anglo-American education in literary art is based on 'canonical' texts. Take the Norton Publishers' *Anthologies* as widely read examples of canonical texts: these anthologies form the basis of a very large number of literature courses at BA degree level. Most of the writers in these anthologies were white, male, Christian and relatively affluent. They were published in the first place and continued to be appreciated, because there was an audience and a growing critical tradition of like-minded people who found and valued in their work the hard labour and aesthetic joy of recognition when an appropriate articulation had been voiced. Look at the selection in Matthew Arnold's first anthology for the teaching of English literature in schools after the Education Acts of 1867 and 1870: the writers include Shakespeare, Pope, Wordsworth and others, but also a large number of texts by his friends and contemporaries such as Browning, Swinburne, Tennyson – and himself. Needless to say, despite the value of these writers to Arnold, very many others, including women, were excluded. Furthermore, because of the extraordinary pressure to legitimise the study of *English* literature, only taught in universities for the first time at the end of the nineteenth century, there were also firm conventions about generically respectable kinds and modes, which excluded diaries, letters, and all but the most politically significant autobiographies.

This history has been reasonably well documented,[16] and has run alongside the feminist revaluation of the canon in the 1970s and 1980s which resulted in more texts by women being republished and put on courses, as well as in the steady, considerable, and serious scholarly work on critical vocabularies for practices of reading diaries, letters and autobiographies. However, if one asks the feminist standpoint question of the literary institution: because of their marginal and excluded position could women do artistic and critical work differently? And if so how? One finds the debates about the way white middle-class feminist theory tended to universalise women's knowledge and exclude the variety of standpoints held outside that institutional group, or about the implications of those specific standpoints for various aesthetic evaluations, but little critique of the still largely masculinist literary institutions. Yet both are necessary.

An exception, among a few others, would be Hill Collins, who brings a concern with both to her analysis of a number of literary texts. She notes in 'Toward an Afrocentric epistemology' (Hill Collins 1990) the legitimation criteria for the 'knowledge validation process' that allow 'Eurocentric masculinist process' to control the credibility of knowledge claims. The process itself is tied to a positivist methodology that requires a distance between the 'subject' and 'object' of study, the 'absence of emotions', the 'inappropriateness' of ethics and values, and the privileging of 'adversarial debate': all clearly related to the rhetoric of the representative democracies outlined by Carol Pateman, and at the centre of the feminist critique of science. Although, as we shall see, Hill Collins offers an alternative in black feminist standpoint methodology, it is not her point to engage in a specific critique of dominant aesthetics. However, as a white, middle-class academic, teaching within that aesthetics and perpetuating its dominance, it is here my point, if not my responsibility, to do so.

An on-going critique of the arts and criticism, of aesthetics, could begin to resolve the temporary hiatus that has come about in the critique of the sciences. A standpoint critique can be taken to arts and criticism very cleanly, since for the past 300–400 years most 'canonical' art has been produced by the 5–20 per cent who have been recognised citizens, subjects of the state. This is largely for economic reasons; citizens were usually by definition propertied, therefore more or less affluent enough to buy the leisure time necessary to work on art; they lived within a community of capitalist owners of the various institutional and industrial technologies needed for the dissemination of the arts: for example the capital-intensive business and industry of publishing and printing which has such a large impact on book production. Even though proximity to this community has slightly less impact on the writing in periodicals, it is only necessary to look at the rationale for relaxing the press controls in the 1840s (that anyone substantial enough to own a press would probably be a self-interested member of the middle classes) to understand the continued influence. These canonical writers address issues pertinent to their lives, which, because they are at least partially represented subjects of the state meant they were addressing represented and representable issues. Hence their work, like that of scientists within their privileged scientific communities, could be appreciated by others in similar positions, with access to the media (and it is media technology that conveys the representations that make ideology possible), and generating a critical vocabulary for discussion of common interests: a critical vocabulary that draws on concepts of (in)adequate language to license the artist as transgressor of ideological representations and provider of a transcendent language.

This critical vocabulary for appreciation is an embodiment of the liberal social contract: the isolated genius who simultaneously can speak on behalf of all, conveying absolute truth, through pure beauty. It is a vocabulary for the subject, and not for the individual writers who appear, from say the letters and diaries of the Romantic poets, to have engaged in quite different relations in the impingement on their writing of the non-ruling areas of civic and domestic life. Furthermore, from a standpoint position, the 80 per cent who are not represented, however inadequately, are not only attempting the strategy of pushing embodiment into ideological representation, but also attempting the strategy of articulating the not-yet-represented or even embodied. These people know there is only a restricted amount that can be gained by worrying about transgression and transcendence, the fruits of (in)adequate language, if one is not represented at all.

There have been attempts to criticise the isolated genius, for example Barthes or Foucault on the 'author', but these are often treated as an erasure of the 'individual'. Similarly there have been attempts to criticise pure truth, particularly in the work of Derrida, although his critique, along with other moral stances, is usually dismissed as caught in an essentialist/relativist dichotomy. Yet there is no recent critique of beauty as something wrested from ideological obscuring into cultural articulation at the moment it loses its power within ruling relations, no analysis of the extraordinary joy it offers at that moment when it still fits so precisely into the structures of social representation, and no critique of the ensuing pleasure and the conditions of its continuance or dissipation. Most of all, there is no critique of why it is so hard to value aesthetic production from those not in the 5–20 per cent, from valued domestic and civic places that raise the issues of class, gender, age, race and ability. This material tends to be called 'popular' yet it is not analysed as a different aesthetic, taking into account different writers, audiences, media. It is often subjected to the same critical analysis as canonical art, and hence always appears to be inadequate to representation, which is tautological since it is not represented, and then held to fail. The material is dismissed as is the entire field of craft work: it is skilled and with tacit knowledge, but since it is not transgressive or transcendent it is not immediately relevant nor can it be appreciated.

PROBLEMS WITH THE RHETORICAL STANCE OF THE ARTS

When you engage with the standpoint question, could women do artistic and critical work differently? and if so how?, several interesting aspects of the humanities come to the surface. In the first place the arts and humanities could be said to exist precisely in order to discuss the difficulties that arise from the impossibility of neutral objectivity, fixed subjectivity and value-free knowledge. From the start of written records, the debates over textuality and interpretation have focused on the different inflexions between absolute and negotiated meaning, and among coding or copying, transcendence and engagement. Prior to and then concurrent with writing, as discussed in Chapter 4, speech as rhetorical orature raised similar questions.[17] One reason that the arts may not currently be such a focus for critique as the sciences is that too many people take its grounds as self-evidently 'good', without discussing the differences outlined above in the discussion of hypertext, between the arts' provision of context simply as a plausible rationalisation, and the provision of context for the material working out of probably-the-best.

Arts and the artist

The arts are also frequently thought of as escapist, and not as seriously to do with 'real life' as science. They are linked in our culture to pleasure and entertainment, and by some to a notion of rather esoteric 'beauty'. Even though these elements of pleasure and beauty are also found in the practice of science they are not usually given much credence, taken if at all almost as by-products of other actions that have more important impact on the actual world. Many people do not even recognise that science can yield these products, and scientists themselves are often shy about the adrenaline surge of competition and the chase, the satisfaction of 'fit'. That these elements of pleasure and beauty could also be cultural by-products of artistic work that negotiates between textuality and lived experience in understanding values from which we can act, is also not often appreciated.

If not escapist, artistic work is often taken as temporary transgression, licensed opposition to specific subject positions allowed by nation-state ideology. In effect the 'arts' become a place where subjects, by definition until the franchises of the twentieth century, people who are white, affluent men, can contest and even produce alternatives to the inadequacies of the representations they are supposed to assume. As contestation and opposition to ideological repression, this work is revolutionary. But, that it is undertaken by autonomous individuals, forces it into anarchic expression. While the revolutionary and anarchic transgression of capitalist nation-state subjectivity is an important reminder of oppression, the action is usually undertaken by and on behalf of a very small percentage of the population, so that its wider value cannot be in terms of the specific oppression it contests, but *how* it does so – the rhetorical strategy at least could be exemplary for a larger audience. The art-for-art's-sake movement at the turn of this century can be read as a recognition of such art's political insignificance in anything but its formal qualities, which display techniques and devices for transgression, rather than solely as a claim to the purity and transcendence of form. However, while art throws into relief the contextual as it trangresses, it rarely clarifies that this context is usually focused through privilege, and that the context has a wider social engagement. In this the avant-garde is structured exactly like the club culture of science.

In western nation states, the arts are also a place where certain privileged men are seen to be *dialogical* rather than positivist and are respected for it. At the same time, as subjects, they are constrained within an ideological system; their dialogical activity is circumscribed and has distinct borderlines. Concurrent with notions of language as (in)adequate to reality gaining circulation during the seventeenth and eighteenth centuries, the logical development of this circumscription leads to the idea that what artists do is 'impossible': when they break the inadequate, they produce beauty, behave like gods, are genius. To repeat: when valued canonical art, in the name of culture, wrests some element from ideology's obscuring and brings it to language and representation, it finds something that seems 'true' because it fits so neatly into the interstices of social life, something that seems to be 'beauty'. The power of such production occludes the fact that we have few ways of recognising, let alone valuing, elements brought to language from outwith ideological representations and their shadows. Discussion of art does not recognise its activity as necessarily limited and concerned with the negotiation and labour of articulation, even though the personal accounts of writers indicate that this is what they do. Rather art, in much critical elaboration of aesthetics, becomes a quest to render the inadequate more adequate, a heroic venture to transcend the frustrations of constant failure. Because the grounds of a necessarily limited language are taken instead as unexamined *a priori*, when one of these grounds is shifted the work appears to have been transcendent or even divine.

Just as in science, the attitude to the grounds of the arts is tied to an understanding *of tacit knowledge*. Science understands the tacit largely as a field of assumptions to which one assents when one enters the institution of technoscience, and it is fascinated by the possibility of the impossible-to-represent tacit. In contrast, the arts work with the impossible-to-represent all the time, but that work can be cast differently through practices based on (in)adequate language and linguistic transcendence, and those based on taking the tacit as a reality that has not yet been worked with in a limited language. Silence that arises from facing the impossible-to-represent can therefore either be an acceptance of inadequacy, or a recognition of inappropriate words that need more collective work.[18] And of course, silence can at any time be a coercive obscuring of grounds. The recent focus on the term *catachresis* in postcolonial theory does distinguish between the coercive and non-coercive, yet does not sufficiently explore the implications of the differences between consensus and corporate within the latter. Significantly, it has been recognised that the distinction between coercive and non-coercive silence cannot be made without an understanding of historical and social context. Similarly, assessing the differences between notions of (in)adequate language, implicitly idealist, and the limited language of textual materialism, cannot be effected without historical and social placing.

Intellectuals, critics and technology

Along the ideology-subject axis that determines the ground for so many canonical writers, it is possible to suggest that there is a specific activity parallel to that of the artist which many societies attribute to the intellectual. When we study the 'arts' we also usually study intellectual commentary on it, or criticism. In the sciences the two activities are closer together. As is fitting for institutions, the members of scientific textual communities are more densely intertextual than those in the sometimes less institutional formations of the arts. The different positioning of criticism is partly a recognition in the arts of the importance of interpretation to communication, yet this activity too has different kinds of practice that need to be located in the social and historical. Critical commentary is necessary to art because it helps to locate and place it. Most helpfully, it is part of the negotiating process of working on articulation. But just as canonical writers tend to come from a specific class and gender position, so do critics. Intellectuals are a group of people important to the state because they provide the structures, the reasons for such change as is necessary to maintain ideology and its market stability. They can do so purposively but also by default, sometimes trying to incorporate challenge to systematic repression yet by offering a vocabulary for it, moving that temporarily visible repression on to further stability. In a sense the intellectual is to the state what the artist is to the individual subject. Just as individuals need representations to operate within ideology, a process we recognise on an immediate basis as 'fashion', so nation states need cultures to represent themselves in a multinational economy. Tourism, which was differently integrated into nation states when they controlled their economies, can increasingly be thought of as national fashion, a commodifying response to the demands of transnational economies demanding stable markets.

If the canonical post-Renaissance artist is the licensed transgressor of subject positions, the intellectual is the critic licensed to expose ideological assumptions. Both become specific historical concepts to answer a set of political and social circumstances. Canonical artists are usually citizens of the state and transgress representations of subjects, only occasionally working on articulations of life excluded from the systematic because by definition so much of their own lives appears already to be represented, even if inadequately. Similarly, intellectuals, especially those who can earn their living as such, are profoundly implicated in ideological systems. When they are not actively justifying those systems, their criticism is usually self-reflexive. All critics should question the assumptions of their grounds, but only to do so is a luxury – a game permitted to those with the leisure to play it because they know they will not really lose power. Although self-reflexivity is an investigation of power, too often, as in ahistorical postmodernism, it is simply concerned with the failure of its own legitimation. I am quite aware that this could be said of the work I am writing.

If technoscience has become a place where common grounds are not sorted out so that science, from the seventeenth century onward, became systematised knowledge, art also acquired another meaning. Art, which had been skill or craft with articulating the real and knowledge of the real, can, in an attempt to legitimate itself, be cast with science as a leisure pursuit for the aristocracy. The aim doubled back and in the new capitalist nation states, legitimated the artist in terms of class and capital, 'conferring copyright'. At the same time as positioning 'art' within class, artistic work became a place to make money. Art may possibly have been accepted as a middle-class and even aristocratic activity, because it could openly demonstrate the strategies for self-commodification into fame and insertion into ideology. But the arts never became as gentlemanly as the sciences because of the built-in counter-activity of transgressing representation.

For the arts, as in the sciences, technology and the media are the profit-making disseminators of representations held to come from autonomous individuals transgressing the norms. That more and more intellectuals receive such dissemination today, may be an indication of how they increasingly function as the recognised artists of nation-state culture rather than the engaged critics of personal work. Technology advances and consolidates the tautological structure of plausible rhetoric in science, while the media do so for the 'arts'. The transgressions of art, the specific ways it contests particular representations, become the legitimate aesthetics of the time, which in turn, with the replication that claims universality, acquire tautological status. Both the modes of representation and the images of representation acquire this tautological status, for even though the textuality of the arts should mean they are more open to question and negotiation, it sometimes happens that just because they appear more open they are in effect less so because people take their potential for analysis and critique for granted.

The arts, or art and criticism, fully understand the work of language as first-order textuality. However, a large proportion of artists and intellectuals with recognised and disseminated work, come from more or less relatively privileged positions that focus that work not on the failure of inadequate language as with the second-order textuality of the communication of science, but on transgressing that inadequacy and attempting greater adequacy, fit, even transcendence. They are rarely in the position of those excluded from representation who feel the need to work on articulation; more often they are subject to desire. Freud, and later Lacan, seem to describe with some accuracy the strategic effects of the small group of men, like themselves, in positions of relative power with respect to the system of ruling state governments in capitalist nations. If modern science sexualised the femininity of 'nature' in specific ways, modern art and specifically here literary text, has conventionally been the place that men go to be 'other', to be what is missing from representation, to find the place of 'fit' and to satisfy desire. Traditionally this is figured by way of the 'muse', who is invariably feminine. Post-renaissance poetics is replete with images of the female muse as reproducer, who has the power variously to impregnate the poet so he becomes pregnant with the poem, to which he then gives birth (sometimes acting as his own midwife), and to whom he is then the parent (often male once more).[19] The desire in the male writer to be 'other' is transparently a desire to be 'the woman', however that is gendered, and casts a specific gloss on the process of representation as reproduction.

Colonising of vocabulary

Yet this poses further problems for a standpoint critique of conventional aesthetics. Unlike the science critique, which uses domination over woman as an insistent metaphor for the relation of science and nature, a critique of the arts has to deal with an image of woman that is so positive that it subsumes and displaces that of men. In what appears to be an endorsement of the feminist standpoint critique of autonomous objectivity, the arts are where men go not only to be dialogic but to be women. Of course the problem is that this 'woman' is a masculinist stereotype of woman the enveloping nurturer. Several critiques of science begin with F. Bacon's appalling metaphor of nature as a woman to be mastered,[20] and contrast it to the nature-as-nurturer image supposedly coming from the preceding period. However, the medieval topos for nature is more usually as 'vicar', the person invested with the authority of God and often figured as a woman. Seeing woman as natural authority asks for a rather different reading of Bacon's metaphor, no less violent, but it also places woman-as-nurturer in the post-Renaissance muse who is the true counterpart to modern science's nature as woman-to-be-conquered.[21] This 'other' is the repressed that the artist re-members through transgression, whether it be narcissistic, neurotic or psychotic. But by Freudian definition, if you are a woman, the repressed you remember is not yours, or indeed you may remember no repressed at all.[22] Such colonising of key terms like 'dialogic' and 'feminine' makes a standpoint critique of the arts particularly elusive – even more markedly so when the work on women's ways of knowing that adds weight to the critique of the sciences, is precisely the kind of knowing that men claim that they do when they work as artists, as well as when so many standpoint theorists concerned with the sciences gesture with hope to the 'arts', which have been primarily defined by and for a small group of privileged men.

Rather more complicated to position is the art that does work with the limitations of the medium, which I understand as the focus of poetics. When that art is studied in detail, it often enacts precisely the negotiations and communication within a group that is attempting to arrive at decisions that will articulate value and instigate action and agency. This after all is the purpose of poetics. For example, again, the Romantic poets: I have no problem with what they were doing with their poetic. They were addressing profound issues of identity, truth and perception; they were rewriting the possibilities of representation by going beyond the (in)adequate concept of language to an engagement with the limitations of language. They wrote a poetics that took people over a hundred years to understand, significantly becoming popular with the enfranchisement of working-class education, and it is a poetics from which I have learned much in my own attempts to speak about women's experience. But of course these groups are also working within a political system of privilege quite different to our own, and dealing with issues and representations appropriate to their positions of class privilege and with agency for themselves. From a standpoint position the poetics is engaged. Yet the rhetoric of that poetics, its moral and ethical interrelation with ruling relations of government and other non-ruling relations of power, needs to be understood to place the poetics within immediate social practices.

If we take the vocabulary of transgression/transcendence, dialogism, and agency, which is used by feminist standpoint theory to criticise science, politics, philosophy and so on, we find that it is at the centre of western aesthetics and appears to justify arts strategies in general. Furthermore, what is not done is a broader analysis within historical context, that looks across partial knowledge to the relation of partial knowledge to the rest of society. Standpoint theorists would never analyse science without looking at the institutional structure that supports it, partly because it is so difficult to do science without an institution and institutional funding. Yet the arts are not perceived to be institutionally based, and so their critique appears to have lacked that impetus to look at social practice. There is little analysis of the imbrication of the arts into state, national and capital interests, and little assessment of the complexity of poetics with regard to the attendant rhetorical context of moral and ethical issues in society. This is another way of saying that beauty is political, that aesthetics is inexorably concerned with morality and ethics, that partial textuality can, like partial knowledge, be dealing either with systems or with the messier interactions with reality, either with adequacy or with the necessary limitations of materiality: it can be systematic and transcendent or it can be situated.

CRITICISM, AESTHETIC VALUE AND READING LITERARY TEXTS: EXAMPLES FROM COURSE WORK

In the arts and its criticism, much recent work has focused implicitly or explicitly on the ideology-subject axis, on analyses and descriptions of ideology, or of the subject, or of the discourse analysis that shuttles between the two. But, as I argued above, there is still no effective critique of the autonomous writer or reader.[23] Throughout the 1970s and 1980s I made such theory a self-conscious focus for my work as a researcher and then as a teacher, because of its power to expose the assumptions of the liberal capitalist societies whose literary products I taught. As I became increasingly interested in writing that grappled with domestic settings, from family relationships to issues of bodily ability, it became apparent to my students and myself that we lacked a critical vocabulary with which to discuss and value these texts. One course in particular, on recent Canadian writing, focused these problems and I will now refer to it in some detail.

The course opened with a lengthy section on the work of three women: Margaret Laurence, Alice Munro and Margaret Atwood. Many of the issues we were discussing related to mother-daughter and other familial relationships. However, when we tried to focus our discussions through available critical vocabularies, we frequently ended in silence. The problem was not enjoyment, for all the students were highly excited by this field of writing that was new and evidently relevant to their lives. They laughed, cried and grieved with the texts. The problem was a vocabulary for an engaged criticism through which we could bring these texts into our day-to-day actions. Rather schematically we read Laurence's *The Diviners*[24] as a deconstructive text. It openly invites the kinds of strategies that take apart social and literary conventions, and as responsible critics we did not merely leave the pieces on the floor but attempted to reconstruct more appropriate strategies from them. Alice Munro's work[25] is less open to such an approach, but provides continual and sophisticated segments of semiological analysis. Her work has an apparently effortless ability to foreground normative behaviour so that it balances on the edge of the unacceptable and generates considerable ethical debate. For Atwood's work[26] we turned to structural and poststructural analysis and studied the acutely perceptive commentary on political narrative that her writing can achieve as it artificialises the apparently natural with surrealist and allegorical devices.

In each case we found that the critical stance allowed us to make important analyses of political power, to follow the way that the writing was transgressive, defamiliarising or subversive of the ideological representations permitted to the familial relationships of the women in the narratives. Yet our criticism became systematised into one or another subject position discursively contesting its ideological representation, and failing to deal with the messy details and denying any value to domestic complexity. There was no sense that those women in the texts or we readers ourselves, women and men, could work on other ways of building relationships. And we had not talked about why we laughed and cried and grieved with the texts.

Another set of texts on this course followed the development of language-focused poets from feminist communities in Canada: writing by among others Nicole Brossard, Daphne Marlatt, Smaro Kamboureli and Gail Scott. Language-focused writing is often difficult to open up because so many students are taught that grammar and syntax are fixed and immovable. So we began to engage with the texts through the debate about the (un)helpfulness of theory to women writing in Canada, which emerged in the late 1980s.[27] To do so, the students had to take on the psychoanalytic discourse that underlay the debate and was actively employed by several of the writers. This discourse, which Lacan elaborated into the Symbolic and Imaginary, for all the complexity of occasional critics such as Luce Irigaray or Jacqueline Rose, or the growing interest in the alternatives offered by the work of writers such as Elspeth Probyn, is still frequently taken to reduce women to absence, loss and lack. Brossard's work, in for example *These Our Mothers* or *Mauve Desert*,[28] attempts to give women's voices different breath, stress and hope. Reading through standard Lacanian discourse renders the language brittle and theoretically too precise. Using it to value Daphne Marlatt's writing[29] allowed us to take the early poem *Steveston* and find sudden silences in the text, silences that after the 1991 publication of *Salvage*, in which she published the edited-out sections on women from the earlier poem, made chilling sense. But while the theoretical approach allowed us to locate the silences and appreciate the exclusion of women, it did not encourage us to talk about the reasons we had committed ourselves to working on these texts in the first place: For what we valued in the texts concerned itself with friendship between women, with an erotics that can inform any sexuality, but is here developed through a lesbian community.

A third area of work on the course concerned writing about social violence against women. Here the writers had little in common and many were new to publication and to literacy; several writers were First Nations women, some of whom came from traditions of complex orature. The first reaction of most students was embarrassment, at the perceived naivety, crudity and ignorance of literary conventions. Indeed, some readers were so embarrassed by the apparent lapses in decorum that they could not read the books in any engaged way; they could not hear the stories about violence. After trying several approaches, we found the most helpful was to look at the genre fiction devices which most of the texts deployed. Among others, we studied Jeanette Armstrong's interweaving of science fiction and romance in *Slash*, Beatrice Culleton's rendition of B-movie dialogue, the shocking frame of Elly Danike's use of pornographic formal elements, Jacqueline Dumas' fairytale structure, and Rose Doiron's shaping through romantic fiction narrative.[30] The strategy retrieved the texts from dismissal, since many of the readers I worked with would not otherwise have read them. But the strategy once more missed the point: it allowed us to discuss the social violence yet then put it safely away in a box. Genre fiction devices are stable structures precisely suited to ideological representation. They are not intended to be transgressive and hence they tended to give the impression that these texts were descriptive and analytical rather than critical, even, that they were not 'literary'. They allowed us to engage only with the descriptions of social violence, not with its agency.

However, with this group of writings, there were still points of embarrassment that could not be analysed away and which the genre fiction analysis did foreground. We were able to choose to go back to these points as most acutely locating the tensions and difficulties around issues of social violence. In doing so, I was repeating a hiatus in my own research, which had come to a halt on a number of these points.[31] For example, in Armstrong's *Slash*, the narrative ends with the birth of a male child, named Marlon, in whom the hopes of the people are placed. The text allows for a messianic reading which many people in my classes find embarrassing because they have been trained to think of contemporary literature as self-consciously ironising or parodying the religious utopia.[32] In addition, with the name 'Marlon' and its echoes of Brando's gesture toward First Nations peoples at the Oscar awards, many of the students who regard the actor as a cynical romantic were encouraged to find the whole image irreparably naive. In the long discussions about this narrative element, we found ourselves debating between a British, largely white and middle-class view of the image as futile and banal in the face of the tragedy of First Nations history, and on the other hand, an insistent portrayal of contemporary First Nations political agency as hopeful and positive. Gradually, through the labour of working through the textual resistance, the discussion became a study of the pressures we were putting on the text to fit the representations we understood. In this and other similar debates the difficulty of these textual elements puts us in the middle of working on the problem, working with the writer on a more appropriate way to articulate the social violence. It was a way that involved us in learning how to value and respect a standpoint outside of the usual relations of ruling power. At the same time it involved us in understanding what Jeanne Perreault understands from the autography of Adrienne Rich's writing, that 'Whiteness...is not merely a fact, but a stand', that we had to articulate for ourselves. Hill Collins, working from white scholar Peggy McIntosh, extends the idea into the claim that whites are taught not to recognise white privilege,[33] and this is precisely what we had to begin to do. This is not an easy process, but engaged textual work offers a site for carrying it out.

In each of these communities of writing, of genre, of canon, of reading, that are here put forward as examples, the legitimate aesthetic approaches for criticism presented the writer as transgressor, a single individual making little impact on the larger structures of ideology, and unseen/unheard by those structures if it did not transgress at all. The reader too became a represented subject, allowed a position by ideology and on whose behalf the writer transgresses. The process left us feeling deprived of agency, unable to discuss our responses or to articulate the things we found difficult to value yet immediate to our lives. The tacit we were trying to put into words was not a set of assumptions, nor the repressed that ideology would rather we forgot – the shadows of representation that silently embody the subject – but it was lived experience that had not been spoken, recognised, legitimised by mimetic repetition, agreed to, valued and acted upon – the tacit that we all have to work on within the limitations of language.

One way of doing this is to build more appropriate representations. Since the beginning of the twentieth century, more and more people have technically had access to ruling power through the 'vote'. Yet as the 1968 events starkly indicated, many people now recognise that they continue to be unrepresented, and are hammering out new ways of recognising, legitimating and accessing political power. Similarly, many people have technically had the opportunity to gain access to aesthetic and critical power, yet still find themselves unrepresented, and are attempting similar legitimating processes. In a liberal democracy that tends to obscure and hide its grounds for action under naturalised and conventional assumptions, writing and reading have traditionally been licensed activities for questioning and challenging those assumptions. But for an enfranchised population with access to education, there is an acute need to address the rather different issues that arise in training large numbers of people in the skills required to evaluate narrative and poetic assumptions. And once more, we do not want a short-term welfare-state solution for reading and writing. Rather than challenge and transgression by a licensed few, there is a need for broad participatory exchange and engagement in shaping the grounds of aesthetic, social and political action.

A STANDPOINT CRITIQUE OF THE ARTS

Most aesthetic critiques have moved on from the value-free scientistic bases of formalism, structuralism and new criticism. Yet there is still an assumption that the context-providing methodology of the arts is, like hypertext, somehow good in itself, whereas it is often trapped somewhere along the ideology-subject axis, forgetting the civic and domestic, forgetting women, the poor, the differently abled, those of different colour, ethnicity and religion. The vocabulary that begins to recognise the forgotten is often unintentionally disparaging: the forgotten are the mimic men, the subaltern, even most brutally the 'abject'.[34] It is not surprising that cultural studies often seems to people outside text-based disciplines to be 'merely' working with words because the ideology-subject axis deals largely only with representations. Anyone who works with texts knows that we deal with the real, but we cannot just say so, we need a vocabulary for it.

Rather than looking only at the causes of ruling power through ideology, or the effects of ruling power in discourse, perhaps we need to look at the negotiations between ruling and non-ruling power and at the rhetorical structures of non-ruling power itself. Sabina Lovibond proceeds from a statement that the recoil from the universalism of 'integrated subjectivity' moves to the 'point of inability to accept *anything* public as capturing the content of one's thoughts or feelings'[35] to where it is no longer acceptable 'for anyone to *represent* anyone else' (69), to insisting that what we value as individuals must pass public appraisal. Echoing the delineation of positivism outlined by Hill Collins (see above, 168) and making a helpful distinction between 'rationality' and the 'ratiofascism' of positivist objectivity, Lovibond says 'The desire to share in meaningful activity is, in other words, logically bound up with the desire for incorporation into a "community of rational beings"...into a community of thinkers in whom we could have moral and epistemic confidence' (72). Continuing with the distinction between coercive norms and agreement to communal norms, an acknowledgement of 'the claim exerted by communal norms of any kind' obliges us to distinguish between the acceptable and unacceptable in thought and feeling, and hence leads to an 'organised self', a 'centred self' who can expect to be 'held to account' and can work in a public discourse. The distinctions enable questions of evaluation; they allow us to attribute meaning and value to 'difference' (73).

This kind of knowledge is analogous to Seyla Benhabib's call, in 'The generalized and the concrete other', for a moral epistemology regulated by universality with a responsibility to concrete materiality.[36] Later in this article she glosses 'universality' as a 'concrete process', not an 'ideal consensus' (274). In contrast, Jane Flax suggests a responsible epistemology that derives from an understanding of justice that can be used to negotiate between an 'inner reality' and the 'signified subject'.[37] Lovibond sees the ideology-subject axis as bifurcating the psyche into a 'real self' and the 'private self' of desires;[38] the private self is tied to integrated subjectivity in public but the real self develops into an organised or centred self in negotiations with a shared public community. I read Flax as differently casting subjectivity as inescapable if we want to act in public so that justice becomes a way the individual 'manages the strain of being simultaneously public and private, alone and in relation to others, desiring and interdependent' (205).[39] For Flax, justice is a process of reconciliation, reciprocity, recognition and judgement that brings together the private and the public in intersubjective citizenship. Yet while Lovibond posits an alternative to the private self of desires, both she and Flax agree on the importance of transforming individual need into public action. In doing so, as Flax says, one recognises 'differences as well as mutuality...one is forced to negotiate with others and to see the boundedness of one's claims as well as one's mutual responsibility for and dependence upon the character of the "we" ' (207).

Distinction and boundary are necessary to the recognition of difference, and a shared public or community is necessary to take responsibility collectively in recognition of both mutuality and difference. Key to shared community is Flax's notion of 'transitional space'. This she brings from object relations theory in psychoanalysis to outline a process of 'symbolisation' which is the 'creative transformation by the individual of what exists independently in shared reality' (204). Fundamentally tied to the process of justice, this transformation works not only between the private and public selves of the individual, but allows groups of people to negotiate public action. In addition, for responsible collective action groups must be made up of individuals capable of desiring justice, and there is a need for 'visible connections between speech, deliberation, empathy and outcomes' (207).

Making these connections lies at the centre of Lorraine Code's development of a storied epistemology that develops responsibility for others through empathy. Despite its potential for manipulation, empathy – and, I would add, story – is a 'nuanced mode of knowing'[40] through which we learn respect for others so that we can engage responsibly with them (87). This awareness of the materiality of rhetorical stance that necessitates constant assessment of whether norms are coercive or communal, authoritative or consensual, may be akin to Lovibond's warning about having both 'reverence and suspicion' when working between universality and materiality. The point for both Code and Lovibond is to learn to treat others as the 'friend'. In Code's words one learns to respect people as a 'second person', not 'third person' individual. Second-person ethics is central to the feminist ethics project because the strength of commmitment that brings one together in both commonality and difference is based on trust and friendship.[41] This is echoed in Lovibond's distinction between treating people as a 'generalised other' in accordance with institutional norms, and as a 'concrete other' where the relation

> is governed by the norms of *equity* and *complementary reciprocity*: each is entitled to expect and to assume from the other forms of behaviour through which the other feels recognised and confirmed as a concrete, individual being with specific needs, talents and capacities.[42]

The concrete other 'signifies the *unthought*, the *unseen*, and the *unheard*' (287) of prescriptive critical theories.

Despite the reservations that are outlined earlier in Chapter 5 concerning Code's deployment of narrative and story as implicitly a good thing, her work is extraordinarily helpful in the way it begins to lay out a vocabulary for textual valuing. She notes that because we cannot 'know' everyone intimately there is the 'cognitive and moral importance of an educated imagination as a way for moral agents to move empathetically beyond instances they have taken the trouble to know well to other apparently related instances' (92–3). Such textual work with narrative is 'responsible cognitive endeavour' (93), that educates individuals in the dangers and corruptions of social stereotyping. Code continues by saying that empathic knowledge is intersubjective and necessarily ambiguous. Drawing on De Beauvoir's terms, Code defines empathy this way because 'its ambiguity is manifested in coming to terms simultaneously with the other's likeness to oneself, and her/his irreducible strangeness, otherness' (141). In resisting 'attempts to merge one's subjectivity with another, or to subsume the other under one's own perspective', and recognising the simultaneity of mutuality and difference, empathic knowledge is, according to De Beauvoir, love. Specifically, it is the love of Plato's philosophic lover in *Phaedrus*, who engages with love as gift always entailing responsibilities, and as possibility for change, and for whom such love is allied with medicine, gardening and a particular kind of engaged writing. As Hilary Rose concludes, 'It is love, as caring respect for both people and nature, that offers an ethic to reshape knowledge, and with it society'.[43]

A CRITIQUE OF THE ARTS FROM THE HISTORY OF RHETORIC

So, how do we learn how to engage in participatory work with texts, to resist the stereotype, the representations of the subject? To read any text and expect more than a reductive duplication, we have to trust it – something the Romantic poets keenly explored. But more than this, we have to position ourselves so that the text can trust us. We need to read it not as a subject but as a friend, not from allowed and representative structures but with an openness of stance that invites the writer and reader both to meet in the text, to open out textuality into articulations of previously unspoken common ground. Arguably the Romantic poets were also trying to achieve this, although within a different emphasis of political and social history. Readers have learned over the past 200 years of critical engagement, different strategies for becoming friends with Romantic poetry. We have considerable empathy for these white, middle-class, Christian poets in England at the turn of the nineteenth century. More immediately, we can recognise in others this ability to be a friend to the text, and we value it within institutional aesthetic systems by calling it originality. But for me there is an urgency to the need to read women's texts, that will be paralleled in other reading communities not valued by conventional aesthetics, and which is unlike learning to read the Romantic poets, valuable though that is. I do not want to wait 200 or even 100 years to learn to be friends with these texts by women. Women have after all been brought up to distrust other women, and we need to work on ways of undoing that teaching. Furthermore, if we do not learn to value this writing it will disappear because it has no legitimating aesthetic; just as, I have no doubt, there is valuable writing from the eighteenth and nineteenth centuries which has been lost because people who had the power to disseminate it did not work on ways of valuing it. The concerns are directly analogous to the issues surrounding the need for a legitimating epistemology for women's participation in the sciences. Yet while what is being pursued there is the articulation of situated knowledges, what I focus on here is the study of the situated textuality that *is* the articulation of those knowledges.

My approach throughout has been to combine standpoint with issues in the history of rhetoric which bring together textualities, knowledges, society and politics. Rhetoric offers among other things a history of the swing between the autocratic and the communal or social, and while conducting an acute analysis of the pros and cons of each, is also concerned with the complexity of on-going negotiations in daily life, and articulates at least one vocabulary for the non-ruling relations of power through the elaboration of the consensus and the corporate. Classical rhetoric is concerned with social context, and distinguishes between the situated and the negotiated on the one hand, and the enclosed and systematic on the other, as different kind of context. In Aristotle the distinction is between the rhetorics of dialectic and science or philosophy, and in Plato between the rhetorics of philosophy and success. Classical rhetoric is also concerned with truth, and distinguishes among the certain, probable and plausible: the certain and the plausible are the domain of the autocrat or demagogue, while the probable is the domain of the orator who is engaged with the audience in working out probably-the-best set of grounds for action. This is truth determined through moral responsiveness and employed within social ethics. And classical rhetoric is concerned with notions of the individual and the group as wielders of power, with the difference between negotiated and represented power: the monarch and the tyrant, the aristocrat and the oligarchy, the constitutional democracy and the popular or populist democracy. Rhetoric has also always included the position of the audience or recipient, technically under the terms ethos, pathos and stance. It recognises that if ethos and pathos are separate then an unequal power distribution can occur, and that stance, which is the situatedness of communication, includes the rhetor, audience and text. Ethos and pathos are the positions of the citizens, perhaps the seedbed for representations of the subject; while stance is the engaged interactive work of rhetor and audience in the textuality of a particular history.

What classical rhetoric is not set up to do is deal with any political or public social activity as a fixed end, although it can describe the way those ends may constrain closed social groups. Hence its classical form has had little effect on ideology in the post-Renaissance world. As elaborated throughout this book and particularly in Chapter 1, ideology technically has the structure of Aristotle's science: it is enclosed, systematic, self-evident and self-justifying. If you enter it as a subject, you assent to the rules. In the classical period it would have been difficult to maintain such a structure for very long, but with the increasingly normalised and extensive media communications of current technology it has become very effective in western liberal democracies. However, as I have suggested, if Aristotle's critique of science is applied to ideology, a highly acute account of political representation can be realised. If that critique, by rhetoric of ideology, is extended to aesthetics, we can derive a workable vocabulary for distinguishing between the subject and the individual in terms of ethos and stance, between objectivity/subjectivity and other argumentation and ordering in terms of logic and dialectic, and between the essentialist/relativist dichotomy and the negotiated in terms of the certain/plausible dichotomy and the probable. Agency and dialogism can be understood not only as transgression and transcendence, but as engagements of moral and ethical negotiation.

What feminist historians of rhetoric have done over the last ten to fifteen years is implicitly to assume a standpoint position and look at those excluded from citizenship to test the applicability of rhetoric to relations of non-ruling power. A rhetorical analysis from the standpoint of women as effectively disenfranchised and excluded from ruling relations of power, would take the 'death of the author' argument, made explicitly through rhetoric by Barthes, and insist that it is not the death of the individual, only the subject. This is something auto(bio)graphy studies do. Within those studies standpoint first insists on personal materiality, the reality of individual experience and existence, and the rhetorical analysis elaborates on the kinds of ethos and pathos, and the effects of stance. In effect this analysis is going on, yet an overt recognition of the rhetorical dimension could extend it beyond the individual into the socio-political. There could not only be analyses of non-ruling relations within a position, but also across positions, and with regard to ruling relations an understanding that the negotiations of the individual are messy and broadly involved, rather than simply caught into discourse systems. With this perception of broad rhetorical strategy, Hill Collins comments that 'the significance of an Afrocentric feminist epistemology may lie in how such an epistemology enriches our understanding of how subordinate groups create knowledge that fosters resistance'.[44] She also extends the concept of personal materiality into rhetorical stance when she insists that a standpoint position not only clarifies the 'objective truth' for a particular group but can also insist on its probability rather than its completeness. She says 'Each group speaks from its own standpoint and shares its own partial, situated knowledge. But because each group perceives its own truth as partial, its knowledge is unfinished' (236). In this, credibility derives from the 'owning' of position rather than its erasure.

A rhetorical analysis from the standpoint of women would take the arguments of Derrida about the insistence of presence and the absolute alongside pluralist relativism, and base the images of fold, pli, seme and so on, within a situated knowledge. Derrida tried to do this through the 1980s, especially in his book on Mandela, yet every position he put himself into was still ideologically privileged. He has rarely, for example, discussed his own racial and cultural background. A feminist standpoint critique would first position itself in an historical immediacy, in order to look at the messy relations among people within non-ruling power, even and especially people without representation, and to look at what so many are now discussing as ethics. A rhetorical analysis would add to this an understanding of morality and ethics as engaged and negotiated best-probable grounds for action, rather than cases of relativism or the absolute, grounds that do not have to stay in a fixed position but whose strategies can cross specific groups and address and change ruling relations.

There are further advantages of bringing together standpoint and rhetoric in both epistemology and aesthetics. Critiques of epistemology within standpoint argue for a critical reality, a critical rationality, a critical objectivity, as they uneasily defend the real while in constant tension with the absolute/relativist divide because they recognise no way of speaking about the probable. The rhetorical analysis of epistemology provides a vocabulary for talking about the negotiated reality, negotiated not on plausible but on probable grounds, grounds worked on by people within a community and across communities through rational, analytical, syllogistic, topical, analogical, symbolic and other logics that necessitate an understanding of the complexity of knowledge and identity in public as neither wholly citizen, subject or private, but in terms of ethos, pathos and stance, in other words identity is not isolated but in relation to other human beings. Probable grounds necessitate an understanding of knowledge about reality as a matter of engagement and negotiation between ethics, or social agreement, and morality, or individual or group agency.

Just so a critique of aesthetics within standpoint would argue first for the need to value the 80 per cent of excluded art, but not through a 'critical' poetics or aesthetics that lead to a philosophical hiatus, but through a rhetorical analysis of aesthetics that offers a vocabulary for talking about the articulation of tacit knowledge by way of a textuality that understands limitations, a situated textuality. The work on words would not be a second-order textuality satisfied with inadequate language and reduced to encoding, nor a first-order textuality continually transgressing inadequate language toward the more adequate by way of transcendent beauty, but a textuality where people work on words together to build common ground. In this attention to common ground, epistemology and aesthetics overlap, as a situated tacit knowledge becomes articulated and therefore textual. An attempt to find ways of teaching and learning about this understanding was the primary impetus behind the development of hypertexts that I described in Chapter 4.

In practice, learning how to be a friend to the text means first of all working on words and opening out textuality into articulations of different and immediately common ground. This work is vulnerable work. Unravelling the representative leaves moments, sometimes longer periods, of freefall. We need support and we need to want to do it, and hence we need communities with a shared sense of what has to be done, if not of urgency. Textual communities for opening out and building common ground usually go on in writing, particularly in the activity before the written enters the commercial world. The communities are often intimate and personal, families, writing/reading groups, newsletters, newspapers, magazines and, recently, e-mail discussion groups. These communities implicitly involve not only writers but also their audiences, the readers. In these settings people work, they labour on the articulation of grounds – often initially with just a glimpse of commonality and shared ground that we seize upon and repeat, and in the to-and-fro of repetition gradually texture the movement into ground, valuing and legitimating it. A version of this activity is referred to in mainstream criticism as intertextuality, a term that is frequently used without its moral action, impact and effect.

Repetition and coherence are the basis for verbal textuality and the literary tools that have been elaborated before and since the start of written records of poetics, hermeneutics and rhetorics. The phonological, grammatical, thematic, topical repetitions are not in themselves good or bad but more or less historically appropriate for situating knowledge within specific material conditions. The 287 or 10 or 1023 ways of negotiating repetition within a textual field that are listed in the innumerable handbooks of rhetoric from the classical Greek period to the eighteenth century, help us to work on common grounds, not only on our own terms but through the text on the terms of a shared community.[45] I would not want to jettison these tools, just as I would not want to jettison rationality or a critical sense of reality, but put them to different uses. Not toward representations that 'fit' ideology, nor toward the anxieties of the subject when that 'fit' does not occur. But toward work on words, and in that work find out what can be said from the tacit fields of women's lives. This is what 'writing the body' has tried to do, and what the new aesthetics of auto(bio)graphy, African-American studies and gay and lesbian theorists, tries to do: work on articulation not representation. Considerable criticism has come from postcolonial studies, and one of my primary commitments, as I have said, is to learning to read and value the domestic. However, I would also like to learn how to read mainstream texts in a different way.

AESTHETICS À LA CARTE AND AESTHETICS IN THE KITCHEN

Textuality, literature and art in general is not only concerned with first-order language as inadequate, and with creating more adequate representations whose transgressive moment of birth is 'beauty' as 'fit': what I call aesthetics à la carte. Textuality is also concerned with distinction, boundary, language as limited, and working on those limits: the hard graft of syntactic, morphemic, semantic and narrative restructuring that always occurs between at least two people, often more, as words, phrases and stories are repeated back and forth across a gap of individual difference till we find we have netted together a workable common ground. The experience of shared common ground in aesthetics à la carte is a shock of infringed selfhood because it is premised on isolated heroic action that someone else has suddenly got hold of: therefore it is a beauty that should only happen once, be rare and unique, somehow less valuable if in common. But working on language as limited is quite different in tenor. Although it needs courage, it has no heroes. There are no guarantees that there will be a result, that the result will be valued by anyone else. This aesthetic shock is precisely that of shared common ground; to pursue my domestic topos, it is aesthetics in the kitchen, a phrase that tells you a lot about my life, starting with the probability that I have one – a kitchen, that is.[46] The shock of shared common ground viewing language as work on limits is a shock of the recognition of common work, of shared value, it has hope written through it as well as joy.

For a concrete example I would turn back to the discussion among Flax, Lovibond and Code that unties the autonomous and isolated individual and releases them into communal interaction. The activity of being a friend to the text is one which I was taught thirty years ago by my teacher of Romantic poetry, but he provided no vocabulary for taking it as a critical ground. Without that vocabulary, when I suggested to one group of students that they learn to be friends with the text, they interpreted it as an invitation to drop all critical understanding and relax into conventional stereotype. Yet as I worked with the different textualities of these three feminist writers, aware that some of the vocabulary, some of the rhythmic pairings, some of the argumentative development was in common as well as differently weighted, the movements between one and another initiated new trajectories as well as a recursiveness that lent density and complexity to the activity.

I read only very recently Marilyn Frye's 'The possibility of feminist theory',[47] and there felt the aesthetic joy of shared common ground. It is akin to finding someone who retains a memory in common with one of your own, and common memory, whether found or worked on, is the basis for all collective action. Frye describes what I have called aesthetics in the kitchen, in an eloquent and emotional register that I value highly yet rarely myself carry out, and which provides me with a confidence to offer the version that I earlier gave and with which I found it had resonance. The quotation is lengthy, but those familiar with it will welcome it again. Frye says:

> women engage in a communication that has aptly been called 'hearing each other into speech'. It is speaking unspoken facts and feelings, unburying the data of our lives. But as the naming occurs, each woman's speech creating context for the other's, the data of our experience reveal patterns both within the experience of one woman and among the experiences of several women. The experiences of each woman and of the women collectively generate a new web of meaning. Our process has been one of discovery, recognising, and creating patterns....Instead of bringing a please of inquiry to closure by summing up what is known, as other ways of generalising do, pattern recognition/construction opens fields of meaning and generates new interpretive possibilities.
>
> (107)

She speaks of blocking expected patterns so that 'In the consequent chaos, they slide, wander, or break into uncharted semantic space' where one can see the things 'denied, veiled, disguised, or hidden by practices and language that embody and protect privileged perceptions and opinions' (108). Pattern perception, similar to what I have called repetition and coherence, includes the ability to be astounded by the 'ordinary', to recognise patterns that do not 'fit', and to press at the 'limits' of pattern (109). In these things we make meanings and form non-homogeneous epistemic communities brought together by story (110).

Frye is describing a process she calls 'consciousness raising',[48] and in doing so is describing a rhetorical activity central to the teaching of participatory and engaged textuality, which many people will recognise as also central to a participatory and engaged politics. The rhetorical strategies help to tap into the personal energy and insight of people and put it to work on articulations of their lives, rather than allowing it as it so frequently can, to become a destructive violence spun off from the exclusion and repression that working within the ideology-subject axis only makes clearer – hence the dislike of so much contemporary discourse theory.

What happened to my recent Canadian Fiction class? Determined by my position in a British academic institution, even if I had read Frye's article, I could not have started consciousness raising groups. Instead, I was lucky enough to work with two communities of people similarly committed to building different ways of situating texts.[49] Through engaging with their activities, I introduced personal diaries and then learning journals as a way of constructively approaching this energy in years three and four of the course. The first year, working only from my own experience, was a qualified success. It must be said that learning journals are rarely if ever used by English departments in the older universities in Britain, nor is group work common. I was told that students usually took English Literature courses precisely so that they would not have to deal with people. Those first students were indeed in some cases made too vulnerable. But the following year I had professional help from Rebecca O'Rourke, a teacher's teacher specialising in creative writing, and the results were warming. The students learned how to work with others, and began to use parts of their lives to understand the texts in an explicit manner. They also began to assess the available literary tools in terms of their relevance to the work rather than allow themselves to be determined by their historical specificity – for example they avoided 'ironising' for the sake of cultural fashion.

The procedure, which is by no means the only way of learning to position oneself so that the text can trust us, did ask the students to read as if the texts were able to help them value aspects of their own lives. By doing so, we were explicitly making a connection to the recent work in autography, which implicitly makes a standpoint critique. At the same time we were also asking the students to be aware of the rhetorical stance by which they were engaging the text. This is not easy. Faced with a seventeenth-century letter for the first time, or with a narrative from Yukon Elder writer Annie Smith, or with a Christina Rossetti sonnet, even making sense of the words, let alone their significance, requires patience and the commitment to read and re-read over a lengthy period of time, each reading bringing possibly different strategies and certainly a different location for the reader. From my teaching over a number of years, it takes at least two weeks of intensive study of a short T. S. Eliot poem several months to ferment before a student new to modernist poetry can begin to understand its devices. In my own reading, I can only recount the experience of working with the letters of Dorothy Moore (1642–5). Several of these letters I put aside as uninteresting only to find over a six-year period that I reclaimed them one by one as I learned how to read them. Jeanne Perreault's elaboration of autography is particularly helpful here in its critique of the work of Adrienne Rich and Patricia Williams. Perreault notes that in Willliams' *The Alchemy of Race and Rights*, the writer reads – not only the written and social world but herself,

Reading herself to us as she has been read by reviewers of articles, student evaluation forms, and workers in shops, Williams constructs these layers of texts and interpretation to make herself into a 'bifurcated reader'.[50]

In doing so, she, like Rich, manifests a number of 'selves' in the text, that work together 'as agent, topic, and performance' (66), in a formality that overwrites any sense of essential identity or confessional 'bios', with individual position appropriate to the issues and to the audience being addressed, being written to or read.

If representation ineluctably deals with (in)adequacy, then it is always implicity idealist, always indicating an adequate wholeness or fixed essence, that can reduce the positioning of self to a determinist identity politics. In contrast, articulation necessitates a rhetorical stance that 'positions' in a different way. It works with the structure of dialogue to ascertain common grounds from which decisions and actions can be taken and made. But prior to this we have to have enough sense of our own voice to begin, enough sense of which community we belong to, to be able to write or say ourselves. We also have to learn strategies for 'hearing' others, reading other positions and places. With these standpoints located, the ensuing dialogue can produce a text that can trust us. We have then to learn about dialogue, or more precisely conversation, to ask why we are engaging in it, what it is we want to act upon – and in this there is no need for all to take up the same standpoint, but there is a need for political commonality.

The extent and complexity of reading skills that are needed should be no surprise, since in our Western European and Anglo-American societies we train people not only to write but to read from the age of five to sixteen at the least. However, a dominant aesthetics based on a small proportion of the population has encouraged both critics to insulate themselves against other disciplinary worlds, and people in other disciplines to assume that literary studies teach nothing of practical skill, merely words. There are many texts I am in the middle of reading at the moment, and the opportunity for discussing them with other readers, although rare in research, is invaluable. Even then, it is extraordinarily hard work to negotiate the many interpretations that skilled readers can derive from written texts, and assess the common ground that respects the writer and engages the reader. In a sense, through the creative writing in learning journals, we were legitimating the students' personal involvement in reading by encouraging them to form the kinds of textual communities enjoyed by writers.[51] In doing so, we were also asking them to engage with rhetorical stance and to situate the textuality.

Hill Collins (1990), following Dorothy Smith, speaks of the importance of 'naming daily life by putting language to everyday experience' (111), so that we learn to value as we learn critically to read our lives. The activity is central to the three elements she sees pervading African-American culture: individual expressiveness, appropriateness of emotion, and capacity for empathy; she herself notes the resemblance to feminist perspectives on 'connected knowing' (216–7). Through Hill Collins' 'ethic of caring' there is explicit reiteration of Code's idea of 'empathy', and of Rose's insistence on the importance of love to responsible power and knowledge. In her own writing, Hill Collins demonstrates the way that 'experiences and ideas that are shared with other members of a group or community which give meaning to individual biographies constitute a...level at which domination is experienced and resisted' (228). Perreault's analysis of autography helps to open out this 'meaning' and alert other readers to the potential for community and agency. Autography, as written not only by Rich, Lorde, Millett and Williams, and discussed in Perreault's book, but also by several others including Bettina Apetheker, whose *Tapestries of Life* was another pioneer in this field, can provide examples of practice for teaching and learning about the situated textuality necessary to situated knowledge.

I suggest that this situated textuality would also be central to the development of any revisioned science. Scientists, however, and those commenting on science, rarely engage with this kind of textuality. An exception would be, say, Fox-Keller's biography of Barbara McClintock, or Donna Haraway's *Modest Witness*, with its extraordinary fluctuation of personal voice and theoretical analysis.[52] Closer to practice is Hilary Rose's positioning of her work on the history of genetics within her experience of her own genetic disorder. In 'Subjectivity and sequences: moving beyond a determining culture', she says that she wants to 'link my experience of living in a body defined by technoscience as flawed/disorderly, with my critique of the new genetics as a determining technoscience'.[53] She goes on to say:

> interrogating personal experience is also a means of making connections. Interrogating that 'me' with her sense of a multiple identity, is also a way of making connection with 'you', that other other, named by this proliferating technoscience as disorderly, diseased, disabled, disempowered and above all determined.

As her autography proceeds, she documents a variety of diagnostic and clinical events for which the account of personal experience helped to open up analysis. For example, as a trained sociologist, Rose recognised that the cold style of the lipid specialist treating both her and her son, laid down a basis of 'compliance not shared care'. Because 'she knew scientists', she could handle the approach, but for her son it made 'compliance with the prescribed therapeutic regime almost impossible'. Later on she discusses the way in which research into the social effects of knowing that one has a genetic disorder is skewed by the 'conversational rule of interviewed subjects...not to make matters unbearable for their listeners'. In these two examples alone, Rose not only situates the knowledge, but articulates the sense of dislocation and disempowerment by way of a tense and wild swing between the first- and third-person 'she' of her genetically disordered body.

This dance between the articulate, agential, knower and the real yet determined body, offers a textuality that invites the reader into the location of the situated knowledge, insists on a fully engaged rhetoric.

DISCUSSION

One of the main problems of a standpoint critique of western aesthetics, as with the gesture made toward the arts by standpoint theorists in other disciplines, is that the rhetorical stance of engaged participation is one that aesthetics has always valued. Yet the dominant aesthetics of the privileged few, so implicated in structures of ruling power and subjectivity, uses strategies and techniques that are not necessarily helpful at this point of enfranchised access to knowledge, power and textuality. Much of the canon and its critical appreciation is concerned with the engagement possible to someone defined as an isolated autonomous individual concerned with the purity of beauty and neutrality of truth. When traditional commentators on computing and science like those discussed in Chapter 3, gesture to the arts, it is at this notion of aesthetics that they are waving. As suggested above, I also suspect that because it is possible for individuals to 'do' art without the institutional framework necessary to disciplines such as law and science, a critique of the larger institutional structure that shows up the corollaries with the critique of institutional science, simply has not been seen as necessary. However, in many areas of epistemology, although this book has focused in an exemplary way on the tacit knowledge of women, there is a political, educational and moral need for much more work in this field.

First-order textuality that takes language as (in)adequate, works within the discursive structure of representations that link and type ideology and the subject even more tightly than pre-Renaissance Biblical typology, although determining a smaller proportion of the population. It is evident that this work is politically effective within nation-state politics. Institutions respond to it because it speaks their language, but I suspect the activity is short-term. When this work is attempted in interdisciplinary settings such as women's studies this limited impact is exacerbated. The methodologies in place in various disciplines may have short-term political effectiveness because they address that disciplinary audience, but they are not so appropriate in the long term either to 'women' or to interdisciplinarity. The textuality that recognises engagement with the limitations of language produces articulations that are often difficult to hear, to see, to feel. Yet these practices and activities hold a potential for resistance and agency not found elsewhere. Nira Yuval-Davis notes the importance for dialogue that can produce individuals who act as political advocates rather than representatives,[54] but I am unsure of how the advocate can make the state hear. It is even more uncertain what can be heard by transnational economics. We do not yet know if there are different rhetorical mechanisms for nation-state ideology and global capitalist representation. However, because the state is so embedded into ideology it will not even know how to initiate action, so I suspect that any change needs to come from groups articulating needs; people need to tell their needs and engage the new rhetoric in a textuality, before the state within global capitalism can respond.

Working on women's texts, doing science or arts differently because I am a woman, will produce articulations that do not enter the conventional field of aesthetic and epistemological value, or that do so but are interesting for other reasons. These are reasons we need to recognise, negotiate and value. In doing so we develop extradisciplinary and interdisciplinary working practices, rhetorical activities that are outwith the ideologically legitimated. This is activity recognised by the institution in its ambivalent response to fields such as women's studies, and we must continue to make it difficult to resist the weight and urgency of these excluded worlds, tacit knowledges and vibrant textualities.

Many scientists, when asked to take their time and value what they do, offer a version of scientific practice that is respectful of and engaged with the actual world of nature. Yet the public communication of science, like the textbook evidence, diverts this understanding into language that repeats the neutral code of second-order textuality. Since the rhetorical stance of that second-order textuality lies hand-in-hand with the stance of the enormously powerful technoscience and the ruling relations of capitalism within nation states, it seems to be the only practice available. Most commentators working on a critique of the sciences have focused on its reductiveness; significantly, those closest to science itself are calling for more emphasis on how to do science differently. When science looks to the arts it sees a public communication of artistic practice that certainly recognises first-order textuality, the contextual field of language, yet one that is under continual pressure to respond to discursive enclosure, to the representations of the ideology-subject axis, simply in order to be heard. Much arts communication may evade reductive encoding but is still using language as a means to an end. Just as technoscience uses the predictability of experimental technique to control the actual and achieve success with it, so most arts communication uses the predicability of specific rhetorical devices to control representations. Historically, modern science has resisted the implicit totalisation of working on words in this manner, and has kept to schematic encodings that may be read not only as reductive but also as obvious indications of partiality, of incompleteness.

However, most people working in the arts when taking their time to value what they do, would offer a version of artistic practice that is respectful of and engaged with the actual world of human beings: note, not with the actual world of language, but with that of other people. The media for communication people use is precisely that, a medium. It makes possible the negotiations and engagements between human beings that body forth/enact a personal or organised or inner self, as well as enacting and articulating the shared public of community, of collective interresponsibilities to sameness and difference. Language is a reality of our daily lives. It can be used as a second-order code, or as a means to an end, or as a medium that links us to, engages us with, articulates and enacts our participation in community. When the feminist standpoint critique of the sciences deconstructs and repositions absolute objectivity as a partial standpoint, and works on a reconstruction of a different objectivity, knowledge and power, the arts could interject that *reconstruction can only be done by turning scientific practice into a medium that encourages articulations about relations among people rather than being an end in itself*. This does not jeopardise the reality of the world upon which science has an impact, but neither does it try to make science better in itself. Instead it asks that we think of science as a medium, and the standpoint critique of the arts offered above would enable us to do science better to engage with other people.

This latter discussion has moved a long way from computing science, but computing is first, foremost and obviously a medium. As contemporary critiques of computing and AI indicate, precisely because it is a new medium, it is throwing forward substantial sets of assumptions about its own textuality. For many years explicitly, although now more implicitly, it has concerned itself centrally with knowledge representation and 'expert' or tacit knowledge. Of particular significance, because it is modelling itself on the physical sciences, the kind of textuality that it illustrates repeats that of the communication of science, and in reinforcing the isolation of the individual, the drive to neutral objectivity and value-free logic, it demonstrates particularly clearly the reductiveness of that encoding. Yet it could act as the textual ground on which scientific practice learns how to become a medium, and through which people engage with others: not just providing 'context' but a rhetorical stance that engages people in negotiating the material conditions of their lives.

The dominant aesthetics of criticism, aesthetics à la carte, underwrites the social representation of both science and the arts. The obsession of AI with exact representation is initiated by the concept of beauty as 'fit'. The perception of science as directed by great scientists discovering truths is a misunderstanding arising from the way science uses language, the way it communicates through second-order language. For computing simply to move the basis of (in)adequate language into first-order representation is to miss the point, to miss the agency of science's work with the real and the difficulty of the labour of articulating its experience. However, aesthetics à la carte seems to legitimate the procedure, and lend to the exponential development of computing and AI, an extraordinary authority over the public understanding of science. What could work is a methodology that engages autography in the computing sciences. This is where I would like to go next with my teaching and learning using hypertexts, but I do not yet know the outcome.

Possibly even sharpened by the focus on representation rather than articulation derived from computing, science has come under sustained criticism from a number of directions. The approaches from standpoint theory and the history of rhetoric studied in the preceding chapters, place in stark relief the ways that the institutions of modern technoscience and the methodologies of its representative strategies, although they mirror the rhetoric of the ideology of the nation state and acquire power, also impoverish the epistemology and deprive it of a moral and ethical agency. I have argued that the critiques of science with their frequent gestures to the arts, in effect underline the impoverishment of current western aesthetics, and that not until we carry out a sustained critique of these aesthetics can we properly address the problems that are perceived in the impoverishment of science. The focus of standpoint on situated knowledge and of rhetoric on situated textuality, can generate a different aesthetics that works with the materially immediate. The stance taken up encourages different strategies for legitimating the grounds from which and on which people make decisions and act, so that they do not assume power and take control over reality, but are empowered by their interaction with reality toward agency. Situated knowledge with situated textuality can be thought of as work on a grammar, a poetic and a rhetoric that encourages people not only to 'write oneself into being' as Nicole Brossard comments,[55] but also to 'hear each other into speech', to net together new common ground and shared values. This would, of course, also be doing politics in a different way.

Notes

1 The ethos of the nation state

1 For example, European writers including British writers, such as L. Althusser, D. Beetham, H.-G. Gadamer, A. Giddens, J. Habermas, J. Pocock, T. Nairn and H. Wainwright, have all engaged with this particular question of how the ruling state displays itself to its public.

2 For totalising approaches to ideology that imply that it is an intrinsic part of every human action, see most of the readings of L. Althusser, 'Ideology and ideological state apparatuses (notes toward an investigation)'. One of the few early exceptions to these readings of Althusser's essay is Macdonell (1986), which also points out that much of M. Foucault's work on discourse is in response to such totalising interpretations of ideology.

3 S. Jarratt 1991, first quotation from 12, second from 27. Jarratt's polemic makes the case within a US context that is rather different but recognisable in Europe.

4 Contemporary commentators such as Marshall McLuhan, (*The Gutenberg Galaxy*) or John Pocock (1971) have documented the links between the graphic media for representing political power both within and outside Europe. The connections have been developed in terms of colonial power by E. Gellner, B. Anderson and S. During; see L. Hunter, 'Ideology as the ethos of the nation state', *Rhetorica*, 1996.

5 N. Machievelli, *The Prince* (New York: Dover, 1992).

6 For one lucid exploration see C. Pateman, *The Problem of Political Obligation: A Critique of Liberal Theory* (Cambridge: Polity, 1985 [1979]) 50.

7 For example there is J. Pocock's analysis in *Politics, Language and Time*, especially 125ff.

8 See D. Beetham, *The Legitimation of Power* (London: Macmillan, 1991) 183.

9 For an interesting account of Hobbes' concept of the political rather than natural cast of contractual and state relations, see C. Pateman, ' "God hath ordained to man a helper": Hobbes, patriarchy and conjugal right', in M. Lyndon Shanley and C. Pateman (eds) *Feminist Interpretations and Political Theory* (Cambridge: Polity, 1991).

10 T. Nairn quotes Marx on the 'Jewish question' for a clear articulation of some of this commentary, from Marx, *Collected Works*, vol. 3, 1843–4, 166; in T. Nairn, *The Break-up of Britain: Crisis and Neo-Nationalism* (London: New Left Books, 1977) 16.

11 Macpherson, quoted by C. Pateman in *The Problem of Political Obligation* (1985) suggests that universal suffrage undermines 'the cohesion of self-interest among those, the propertied classes, who chose the representatives' (51).

12 Although this is certainly a focus for T. Nairn, as it had been for Hannah Arendt and before her, Rosa Luxemburg; H. Wainwright points this out in one of the recent works on political theory that is entirely concerned with the issue, *Arguments for a New Left: Answering the Free Market Right* (Oxford: Blackwell, 1994).

13 For political theory commenting on this strategy, see N. Poulantzas, *State, Power, Socialism*, trans. P. Camiller (London: New Left Books, 1978); A. Giddens, *A Contemporary Critique of Historical Materialism: Vol. 1, Power, Property and the State* (London: Macmillan, 1981); and J. Kellas, *The Politics of Nationalism and Ethnicity* (London: Macmillan, 1991).

14 T. Strong, 'How to write scripture: words, authority, and politics in Thomas Hobbes', *Critical Inquiry*, 1993, autumn, 128–59, argues convincingly that as early as the 1640s Hobbes was aware of this strategic need, saying that for Hobbes 'politics must be art so that the public realm confronts us without raising the question of its authorization and authorship, of its authority' (151). For more detail on the connection of this strategy to structures and techniques of the genre of fantasy, see L. Hunter, *Modern Allegory and Fantasy* (London: Macmillan, 1991).

15 See C. Pateman on Hobbes' *Leviathan*, which states that when men form a body politic 'their unity is represented in a very literal sense by the person of their (absolute) master and ruler, Leviathan [i.e. the state]', in ' "God Hath Ordained..." ' (68).

16 See A. Giddens, *Power, Property and the State*, where he argues that with capitalism, the state replaces the city as the 'power container' by offering a 'unification of an administrative apparatus whose power, stretches over precisely defined territorial bounds' (13).

17 See also A. Giddens, *Power, Property and the State*, 5.

18 H. Wainwright, *Arguments for a New Left*, 267ff., in which the analysis of current neo-liberalism also remains uneasy with this argument.

19 The inability to understand this is part of Wainwright's critique of Hayek's neo-liberalism, which cannot for example recognise that 'free' markets are frequently controlled by oligopolies of capitalist competition.

20 Apart from the works cited in Wainwright, above, a helpful study of Adam Smith on this particular point is V. Brown, *Adam Smith's Discourse: Canonicity, Commerce and Conscience* (London: Routledge, 1994) which also notes the radical difference between Aristotle's 'citizen' and Smith's notion of the state that this paper makes (210); the book also gestures to the erasure of individual political 'personality', but does not suggest the development of the personality of the 'state' that is indicated here by 'ideology'. Brown's book generously delineates the recent work on Adam Smith from which I have drawn. For a thought-provoking study that goes to the centre of the matter, see C. Miller, 'The *polis* as rhetorical community', *Rhetorica*, XI:3, summer 1993.

21 A. Giddens, *The Class Structure of the Advanced Societies* (London: Hutchinson, 1980 [1973]), 285–98.

22 See H. Wainwright (1994), who offers the example of 'Fordism' rather later in the 1920s, as the key development of the stability of the US government in the post-World War II years, in *Arguments for a New Left*, where she posits that the ability of management to break down skilled labour into codified parts that unskilled workers could carry out, brought about industrial development with 'high initial or fixed costs' that required 'the mass consumption of standardised goods' to become profitable' (70). See also L. Faigley, *Fragments of Rationality: Postmodernity and the Subject of Composition* (London: University of Pittsburgh Press, 1992) for a pointed commentary on Fordism and postmodernity. His analysis of the production of student subjects through composition textbooks is a more thorough but parallel study to the production

of student subjects by artificial intelligence textbooks that follows in Chapter 2 of this book.

23   N. Yuval-Davis, *Gender and Nation*, 11.
24   D. Beetham, *The Legitimation of Power*, 221.
25   Thanks to H. Rose, to whose many revealing commentaries on modern politics and society I am indebted.
26   D. Beetham, as above, 211–21; or P. Connerton, *How Societies Remember* (Cambridge: Cambridge University Press, 1989) 1.
27   T. Nairn, *The Break-up of Britain*, 20.
28   D. Beetham, *The Legitimation of Power*, 183.
29   T. Nairn, *The Break-up of Britain*, 24.
30   D. Beetham, *The Legitimation of Power*, 107.
31   H. Wainwright, *Arguments for a New Left*, 69.
32   T. Eagleton, 'Nationalism, irony and commitment', in *Nationalism, Colonialism and Literature*, intro. S. Deane (Minneapolis MN: University of Minnesota Press, 1990) 27.
33   John Pocock offers an analysis of early Chinese government which ties legitimacy into print as one of the earliest technologies attempting invariable repetition, where 'li' is the ritual performance of 'truth', and 'fa' is a lesser truth interpreted from writing. The lesser truth is always predicated on the greater, so the written edicts that controlled the large state of China under Han Fei were implicitly involved in evasion, omission, a necessary incompletion that made it impossible to question the law because the edict was already an interpretation of something that was 'true', and was therefore not open to interpretation itself. In a sense, this is to make the act of hiding, of naturalisation, functional to the rhetoric. See J. Pocock, *Politics, Language and Time*, 67ff. Of course, Edward Said in *Orientalism* and other writers have discussed the strategy at length, but not with relation to the rhetoric of nation-state ideology and its ethical topos.
34   E. Gellner, 'Nationalism and the two forms of cohesion in complex societies', 167.
35   See L. Hunter, *Rhetorical Stance in Modern Literature* (London: Macmillan, 1984) 15–18; and R. Scanlon, 'Adolph Hitler and the technique of mass brainwashing', in *Rhetorical Idiom: Essays in Rhetoric, Oratory, Language and Drama* (New York: Russell and Russell, 1966/1958).
36   Another way of thinking about the relationship comes from concepts of the contractarian state, in which the 'legitimation' of the private depends on accepting the determining nature of the public; see V. Medina, *Social Contract Theories: Political obligation or anarchy* (Savage MD: Rowman and Littlefield Publishers, 1990) 153.
37   For one of many commentaries on this phenonmenon, see C. Pateman, ' "God hath ordained…" ', 65.
38   The procedures for 'forgetting' are outlined in G. Orwell's *Nineteen Eighty-Four* and discussed in L. Hunter, *George Orwell: The Search for a Voice* (London: Macmillan, 1984); forgetting is analysed at much greater length in L. Hunter, *Outsider Notes: Feminist approaches to the Nation State, Writers/Readers and Publishing in Canada 1960–95* (Vancouver: Talonbooks, 1995).
39   J. Pocock, *Politics, Language and Time*, 237.
40   See especially the essay by D. Cannadine, but also all contributions to E. Hobsbawm and T. Ranger (eds) *The Invention of Tradition* (1983).
41   P. Connerton, *How Societies Remember*, 63ff.
42   See N. Yuval-Davis, *Gender and Nation*, 14, on the way Foucauldian 'cultures' shift the idea of unitary reification toward interaction at points of resistance on the cultural grid.
43   D. Macdonell, *Theories of Discourse*, 28–9.

44 The question has become a fundamental point of inquiry for feminist theory; see for example *Feminism and Foucault: Reflections on Resistance*, eds I. Diamond and L. Quinby (Boston: Northeastern University Press, 1988).

45 See C. Williams on the effects of repression on the formation of the subject in 'Feminism, subjectivity and psychoanalysis: toward a (corpo)real knowledge', in Lennon and Whitford, 166.

46 See P. Hill Collins, *Black Feminist Thought* (London: Routledge, 1990) 229, where she notes that 'Domination operates by seducing, pressuring, or forcing African-American women and members of subordinated groups to replace individual and cultural ways of knowing with the dominant group's specialized thought'.

47 R. Braidotti speaks of the unconscious as a modern dilemma that marks the division between the subject and its consciousness in 'On the female feminist subject, or: from "she-self" to "she-other" ', in Bock and James, 182. Braidotti, however, suggests that a unitary subject had been in place from the classical period until the modern dilemma, while I would emphasise that I do not think that this is the case, in fact I think the idea of a unitary self is an artefact of the kind of splitting that ideology effects.

48 See for example, T. DeLauretis, *Alice Doesn't: Feminism, Semiotics, Cinema* (Bloomington: Indiana University Press, 1984) 134.

49 See for example, H. Longino, 'Subjects, power, and knowledge: description and prescription in feminist philosophies of science', in E. Fox-Keller and H. Longino, 270. See also J. Henriquez *et al.* (eds) *Changing the Subject: Psychology, Social Regulation and Subjectivity* (London: Methuen, 1984).

50 W. Rowe and V. Schelling, *Memory and Modernity: Popular Culture in Latin America* (London: Verso, 1991) 229.

51 For an extended discussion of the interaction between memory and articulation see L. Hunter, *Outsider Notes* (Vancouver: Talonbooks, 1995).

52 As above, xii.

53 D. Macdonell in *Theories of Discourse*, 35ff., locates this rather differently in Althusser's notion of ruling ideology and subordinate ideologies.

54 J. Butler, *Bodies that Matter: On the Discursive Limits of 'Sex'* (London: Routledge, 1993) 187.

55 One finds a similar neglect of the rhetoric of state power in studies of political ethos which focus on individual rhetoric. The helpful analyses offered by Sullivan and Reynolds, cited above, indicate the presence of state power but no detail about its rhetoric strategies; this can be misleading since the analysis can be carried out implicitly dependent on descriptions of governing rhetoric that derive from state formations peculiar to other times and places.

56 D. Beetham, *The Legitimation of Power*, 240.

57 D. Smith, *The Everyday World as Problematic* (Boston: Northeastern University Press, 1987) 5.

58 For a comprehensive introduction to this now large field, see H. Rose, *Love, Power and Knowledge* (London: Polity, 1994).

59 For a survey of recent commentaries from philosophy of science see Chapters 3 and 5.

60 See H. Rose, *Love, Power and Knowlege*, for an elaboration of this concept. It will also be discussed in Chapters 5 and 6.

61 These are the predominant fields of the women working on standpoint whose work is discussed in the previous chapter. There are of course many more.

2 Rhetoric and artificial intelligence

1 This ground has been covered from different points of view by W. Ong, *Rhetoric, Romance and Technology: Studies in the Interaction of Expression and*

Culture (New York: Cornell, 1971); and B. Vickers, see the article in this collection.

2 The classical education in public schools has been a central location for the transmission of rhetorical skills to a particular class, from the seventeenth century to the twentieth.

3 A. Grafton and L. Jardine, *From Humanism to the Humanities* (London: Duckworth, 1986).

4 See in particular the work on humanist educationalists by L. Jardine, as above, N. Streuver 'Lorenzo Valla: humanist rhetoric and the critique of the classical language of morality', in J. J. Murphy (ed.) *Renaissance Eloquence* (London: University of California Press, 1983); see also the commentary offered in Chapter 4 below.

5 This is an attitude to rhetoric that has recently been promoted in particular by the Frankfurt School studies on the strategies of the mass media.

6 For a slightly extended account see the summary in L. Hunter, *Rhetorical Stance in Modern Literature* (London: Macmillan, 1984).

7 For other versions of this view, see S. Woolgar, 'The ideology of representation and the role of the agent', in H. Lawson and L. Appignanesi (eds) *Dismantling Truth: Reality in the Post-modern World* (London: 1989). A rhetorical commentary on Woolgar's writing may be found in J. McGuire and T. Melia, 'Some cautionary strictures on the writing of the rhetoric of science', *Rhetorica*, VII, 1. For a solid description of practical science writing see C. Bazerman, *Shaping Written Knowledge: The Genre and Activity of the Experimental Article in Science* (Madison WI: University of Wisconsin Press, 1988).

8 See L. Hunter, 'Watson and McLuhan's *From Cliché to Archetype*', 1991.

9 See L. Hunter, *Rhetorical Stance in Modern Literature*, 1984.

10 Plato, *Gorgias* (Penguin, 1960).

11 Plato, *Phaedrus* (Penguin, 1973).

12 Reading Aristotle and Plato as concerned with rhetoric yields a rather different approach to their concepts of knowledge than that gained from reading them within their self-defined community of philosophy. The latter reading has been dominant until quite recently, and underwrites the notion of idealist Platonism which is particular to a group of post-medieval readings of Plato and is not the whole story. For a clear example of the latter see G. Lloyd, *The Man of Reason* (London: Methuen, 1984) Chapter 1.

13 Aristotle, *Topica*, J. Barnes (ed.) *The Complete Works of Aristotle, Revised Oxford Translation*, Bollingen Series LXXI vols 1 and 2, Princeton NJ: Princeton University Press, 1984. See also the helpful distinction made by E. Garver, 'Aristotle's *Rhetoric* on unintentionally hitting the principles of the sciences', *Rhetorica* VI: 4 (autumn, 1988).

14 This use of 'truth' puts to one side the possibility of a 'true' as 'reliable' or 'dependable', and focuses on 'true' being 'as the case is', exact and accurate.

15 See for example Bacon's discussion of the 'idols of the tribe', or his specific comments on the role of rhetoric in *The Advancement of Learning*.

16 See for example some of the discussions of pedagogical technique in H. Jahnke and M. Otte (eds) *Epistemological and Social Problems of the Science in the Early Nineteenth Century* (London, 1981).

17 T. Sprat, *The History of the Royal Society of London, for the Improving of Natural Knowledge* (London: J. Martyn, 1667).

18 This kind of separation is still very much with us. See P. Medawar, 'Is the scientific paper a fraud?', in J. Brown *et al.* (eds) *Science in Schools* (Milton Keynes, 1986 [1964]).

19 For example, I followed Ong on this approach (see L. Hunter 1984) until quite recently; and many writers on the discourse of science support this view.

20 For example, Heisenberg's proposal of 'uncertainty relations'.

21 T. Kuhn, *The Structure of Scientific Revolutions* (Chicago IL: University of Chicago Press, 1962).

22 L. Faigley, *Fragments of Rationality: Postmodernity and the Subject of Composition* (London: University of Pittsburgh Press, 1992) 23.

23 P. France offered this overview in the Durham University series on Rhetoric, 1992, later collected into R. Roberts and J. Goode (eds) *The Rediscovery of Rhetoric* (London: Bristol Classical Press, 1993).

24 Indeed the chronology has been reconstructed for cognitive psychology and artificial intelligence (Haugeland 1985).

25 J. Searle, in 1980, proposed the 'Chinese-room' experiment to counter some of the implications of the Turing test in 'Minds, brains and programs', *Behavioural and Brain Sciences*, 3. This debate has been pursued by a number of writers, notably S. Harnad (1989) 'Minds, machines and Searle', *Journal of Experimental and Theoretical Artificial Intelligence*, I; and by Searle himself and P. M. and P. S. Churchland in *Scientific American*, I, 262 (1990).

26 Most participants in this debate agree that any resolution of the debate is so far in the future that current argument must be purely speculative. I personally read the conceptual debate as a red herring, although the ethical debate over the social and political (and biological) rights of technological objects that behave like humans is interesting and of course provides much of the subject matter for contemporary science fiction and fantasy.

27 Haugeland 1985: 4.

28 T. O'Shea, 'IKBS – setting the scene', in T. O'Shea *et al.* (eds) *Intelligent Knowledge-based Systems: An Introduction* (London, 1987) 11.

29 The predicate calculus and formal logic from the cornerstone of every artificial intelligence textbook, even for example the more generalist *Principles of Artificial Intelligence* by N. Nilsson (New York, 1982).

30 N. Nilsson, as above, 145.

31 Interestingly, this is also the basis of the rhetorical theory of argumentation put forward by C. Perelman and L. Olbrechts-Tyteca, *The New Rhetoric: A Treatise on Argumentation*, trans. J. Wilkinson and P. Weaver (London, 1971). The theory is open to many of the same drawbacks and extensions of the artificial intelligence approach.

32 As outlined in J. Huizinga, *Homo Ludens: A Study of the Play-elements in Culture* (London, 1949 [1944]).

33 Weizenbaum 1976: 14.

34 Weizenbaum 1976: 115; see also J. Ellul, 'The power of technique and the ethics of non-power', in Woodward 1980: 246, for a discussion of the evasive power of the computer technology, and T. Roszak, *The Cult of Information: The Folklore of Computing and the True Art of Thinking* (Lutterworth Press, 1986) 39.

35 In fact Weizenbaum says 'man', not 'human beings'. This discourse of AI is repetitively sex-specific; one questions the significance of the exclusion. See also J. Bolter, *Turing's Men: Western Culture in the Computer Age* (University of North Carolina Press, 1984) 13, for comments on this fear.

36 Haugeland 1985.

37 Haugeland 1985: 197; H. Dreyfus, 'From micro-worlds to knowledge representation: AI at an impasse', reprinted from 1979, in J. Haugeland (ed.) *Mind Design: Philosophy, Psychology, Artificial Intelligence* (London, 1981) 201.

38 L. Wittgenstein, *The Blue and Brown Books* (Oxford, 1952).

39 I am most grateful to Lesley Jeffries for discussions on these issues.

40 Quoted in J. Ennals (ed.) *Artificial Intelligence, State of the Art Report*, 15:3 (Oxford, 1987) 152.

41 See the section on 'Natural language' in M. Sharples *et al.* (eds) *Computers and Thought: A Practical Introduction to Artificial Intelligence* (London, 1989).

42 M. Minsky, 'A framework for presenting knowledge', [reprinted from 1975], in Haugeland 1981.

43 D. Waltz, in 'The prospects for building truly intelligent machines', *The Artificial Intelligence Debate* (Graubard 1988) suggests that new procedures for associative memory 'can provide a very powerful heuristic method for jumping to conclusions, while traditional AI can be used to verify or disconfirm such conclusions' (200).

44 R. Turner, *Logics for Artificial Intelligence* (Chicester, 1984).

45 E. C. Way, *Knowledge Representation and Metaphor* (Oxford, 1994) gives a useful study of the dominance of symbolic logic in natural language programming, while remaining committed to the idea of type hierarchy as 'neutral' (241).

46 W. Black, *Intelligent Knowledge-based Systems* (Wokingham, 1986/7) x.

47 D. McDermott, 'Artificial intelligence meets natural stupidity', reprinted from 1976, in Haugeland 1981: 154.

48 J. Fodor, 'Methodological solipsism considered as a research strategy in cognitive psychology', reprinted from 1980, in Haugeland 1981.

49 Dreyfus 1981: 197.

50 Weizenbaum's example is of mother–child bonding, which is culturally significant but rather detracts from this physiological argument.

51 Turkle 1988.

52 Cowan and Sharp 1988: 113.

53 Dreyfus and Dreyfus 1988.

54 P. Winston, *Artificial Intelligence* (London, 1984 [1977]).

55 N. Nilsson, as above, 17.

56 D. Michie and R. Johnston, *The Creative Computer: Machine Intelligence and Human Knowledge* (Penguin, 1985 [1984]).

57 E. Charniak and D. McDermott, *Introduction to Artificial Intelligence* (Wokingham, 1985).

58 Ong 1971.

59 See for example the alphabetic breakdown in bp nichol, *The Martyrology* (Toronto, from 1976) and R. Kroetsch, *The Sad Phoenician* (Toronto, 1979); Canadian literature is particularly rich in graphical and typographical experiment.

60 What do they make of W. Wordsworth's 'half create and half perceive' or J. Keats' 'negative capability', one wonders?

61 Examples run throughout the book but see in particular 13, 94, 214.

62 See J. Turner, as above, 18; or D. Dennett, 'Where am I?', in D. Hofstadter and D. Dennett (eds) (1981) *The Mind's I: Fantasies and Reflections on Self and Soul*.

63 Textbooks from the the late 1980s frequently specialise and focus in on one aspect or another of AI, indicating the growing complexity of the field.

64 Charniak and McDermott 1985: 6.

65 For just one among many recent misunderstandings of the process of writing, that in this instance claims that print did not change the structure of the written, see R. Sokolowski, 'Natural and artificial intelligence', in *The Artificial Intelligence Debate* (Graubard 1988).

66 L. Hunter, 'The computer as machine: friend or foe?', in C. R. R. Turk, *Humanities Computing* (Chapman and Hall, 1990).

67 P. Gaskell, *A New Introduction to Bibliography* (Oxford: Clarendon Press, 1972).

68 The way in which I use the term 'data' is set out earlier in this chapter. However, readers should beware that people within the computing science community use the term in many different ways. The use of it here has more in common with that outlined in Charniak and McDermott (1985: 319–20) than for example in

C. Date, *An Introduction to Database Systems* (Wokingham: Addison-Wesley, 1986) 5.

69 For example see the study of matrices written in the aftermath of structuralism by M. Foucault, *The Order of Things* (London: Tavistock, 1970).

70 For a brief account of some factors surrounding the early definition of rational analysis in modern science, see L. Hunter, *Rhetorical Stance in Modern Literature* (London: Macmillan, 1984).

71 For an account of the systematic structure of fantasy narrative see L. Hunter, *Modern Allegory and Fantasy* (London: Macmillan, 1989).

72 There are several ways of reconstructing such knowledge; for example see the excursion of J. Lyotard and G. Leech in L. Hunter, 'The computer as machine: friend or foe?', in C. C. R. Turk (ed.) *Humanities Computing* (London: Kogan Page, 1990).

73 See for example the background chapter in S. Deen, *Principles and Practice of Database Systems* (London: Macmillan, 1985).

74 See for example, a classic discussion of similar issues in A. Cicourel, *Method and Management in Sociology* (Macmillan, 1964).

75 See R. Jakobson, 'Two aspects of language and two types of aphasic disturbances', in R. Jakobson and M. Halle, *Fundamentals of Language* (The Hague: Mouton, 1956); G. Genette, *Figures* (Paris: Editions du Seuil, 1966); W. Booth, *A Rhetoric of Irony* (Chicago IL: University of Chicago Press, 1974); and M. Halliday and R. Hasan, *Cohesion in English* (New York: Longmans, 1976).

76 A. Moreton, *Literary Detection* (Scribner, 1978).

77 During a visiting lecture given by Professor Burrows in October 1988 at the University of Leeds, he displayed several studies of the chronological shifts in the vocabulary of common words in historical novels from 1750 to 1950, indicating that genre may be read historically in terms of stylistic consistencies.

78 The history of a main tool in this field, the Oxford Concordance Program, is a good example of both the profound intellectual challenges that must be faced as well as the pragmatic restrictions of computer technology in the 1970s. See S. Hockey and J. Martin, *Oxford Concordance Program* (Oxford University Computing Service, 1988).

79 Artificial intelligence textbooks usually have a section on 'natural language' problems; see for example Charniak and McDermott (1985).

80 See for example, E. Way, *Knowledge Representation and Metaphor* (Oxford: Intellect Books, 1994).

81 Professor Burrows was kind enough to allow the School of English at Leeds access to his editions.

82 I owe a considerable debt to Mr. T. T. L. Davidson of the Linguistics and Phonetics department at the University of Leeds, for discussions on the mark-up of the Burrows' editions.

83 See J. Burrows, *Computation into Criticism* (Oxford: Clarendon Press, 1986).

84 S. Urkowitz, *Shakespeare's Revision of King Lear* (Princeton University Press, 1980).

85 G. Cook, *Discourse* (Oxford: Oxford University Press, 1989).

86 Many computing scientists would dispute whether this were a database at all because there is so much duplication of information. For a sound introduction to the field see C. Date, *An Introduction to Database Systems* (Wokingham: Addison-Wesley, 1986).

87 The printed versions of these databases may be found in D. Attar, *Household Books Published in Britain, 1800–1914* (London: Prospect Books, 1987) and E. Driver, *Cookery Books Published in Britain, 1875–1914* (London: Prospect Books, 1989).

88 See L. Burnard, *Famulus Users' Manual* (Oxford University Computing Service, 1985).

89 EXTRACT is a programme developed at the University of Leeds by J. Duke and T. Screeton of the University Computing Service.

90 For information about the early ESTC, see the outline by M. Crump in *Factotum*, 'Searching ESTC on BLAISE-LINE', Occasional Paper 6, January 1989.

91 For example, see T. O'Shea, 'Machine learning', in O'Shea *et al.* 1987, which presents a structural analysis of plot in Shakespeare's *Macbeth*.

92 This is carried out through a process computing scientists call 'normalisation'.

93 J. Burrows, 'Modal verbs and moral principles', *JALLC*, 1:1 (1986).

94 This is also the classic formulation of the difference between short-term and long-term persuasion in the field of rhetoric.

95 See S. Stigler, *The History of Statistics* (London: Belknap Press, 1986).

96 The co-occurence of the mathematical and the visual stress is interestingly presented in Haugeland 1985.

97 T. Merriam presented these views in a conference paper given in Edinburgh in 1984, and kindly provided me with a copy.

98 J. Burrows, in 'Word patterns and story shapes: the statistical analysis of narrative style', *JALLC*, 2 (1987), uses eigenvalue statistics to generate these designs.

99 B. Erickson and T. Nosanchuk, *Understanding Data* (Open University Press, 1983).

100 It cannot be emphasised strongly enough that rhetorical stance, an interactive and social process, has nothing to do with D. Dennett's 'intentional stance'. Dennett's interpretations of philosophy, recently in 'When philosophers encounter artificial intelligence', in Graubard 1988, are quite specifically within rhetorical strategy. The result is here, for example, a troubling reading of the birds in Plato's *Theaetetus* as a question of possession and command, rather than a reference to the difference between fixed and contextualised data, or information versus knowledge.

101 D. Hofstadter is moving toward this in his commentary on Zen koans, 'Human and Gödel', in *Gödel, Escher, Bach: An Eternal Golden Braid* (Penguin, 1979); however, he fuses dualistic logic with words in order to explain the activity of the koan writing, which leaves him stating that 'a major part of Zen is the fight against reliance on words' because they are limited and fail, rather than an alternative construction in which the necessary inadequacy of words is engaged with in order to indicate the limitations of dualistic logic.

102 This is an aspect that Michie and Johnston admire, as above, 195.

103 J. McCarthy, in 'Mathematical logic in artificial intelligence' (in Graubard 1988) approaches this attitude when he says that AI may lead to a 'study of the relation between a knower's rules for accepting evidence and a world in which he is embedded' (310).

3 AI and representation

1 D. Gooding, ' "Magnetic curves" and the magnetic field: experimentation and representation in the history of a theory', in Gooding *et al.*, 1989, 218.

2 Quote from D. Lenat and R. Guha, *Building Large Knowledge-based Systems* (Reading MA: Addison-Wesley, 1989).

3 J-C. Gardin, 'Interpretation in the humanities: some thoughts on the third way', in Ennals and Gardin 1990.

4 See the recent studies by J. Swearingen (1991), S. Jarratt (1991) and S. Poulakos (1995), which offer the pre-Platonic sophists as examples of rhetors concerned with multiple voices. Within this context Plato can be recast not as rigidly

authoritarian, but as a writer concerned with how to make the decisions upon common grounds that are necessary before any social action can be taken.

5 H.-G. Gadamer, *Reason in the Age of Science*, 1976.

6 J. Habermas, *The Philosophical Discourse of Modernity* (Polity, 1987).

7 See for example the analysis of the rhetoric of totalitarianism and authoritarianism in the work on nationalism by Gellner (1982), Hunter (1984) and Rowe and Schelling (1991).

8 For arguments in favour of club culture see R. Rorty, *Objectivity, Relativism and Truth* (Cambridge: Cambridge University Press, 1991) and *Essays on Heidegger and Others* (Cambridge: Cambridge University Press, 1991), and for various critical, if not condemnatory, arguments see F. Jameson, 'Foreword' in Lyotard (1984), and Z. Bauman, *Modernity and Ambivalence* (Polity, 1989) and *Intimations of Postmodernity* (London: Routledge, 1992).

9 J. Lyotard, *The Postmodern Condition: A Report on Knowledge*, trans. G. Bennington and B. Massumi, fwd by F. Jameson (Manchester: Manchester University Press, 1986 [1979]).

10 J. Habermas, *The Philosophical Discourse of Modernity*, 194.

11 G. Geiss and N. Viswanathan, *The Human Edge: Information Technology and Helping People* (New York: The Haworth Press, 1986) xxi; and S. Woolgar, 'The turn to technology in social studies of science', *Science, Technology and Human Values*, 16, 1 (winter, 1991).

12 Whether or not it is historically accurate to portray pre-Platonic sophistic rhetoric as concerned with an unstable ethos, recent interest in it indicates the anxiety about political structures that have to deal with multicultural, multiracial and multiclass diversity; see above, note 4.

13 Much of this work over the past thirty years has been focused on the writing of J. Derrida, who problematises 'ethos' in 'Plato's pharmacy' (1967) and could be said to have extended the notion of deconstruction specifically to deal with the problems raised by the inadequacy of ethos for an increasingly varied audience.

14 See the classic formulation of the way technology and experiment are part of reality in P. Bachelard, *La Formation de l'Esprit Scientifique* (Paris: Vrin, 1938).

15 D. Kirsch, 'Foundations of AI: the big issues', *Artificial Intelligence*, 47 (1991), 5.

16 See T. Adorno, *Against Epistemology: A Metacritique* (Oxford: Blackwell, 1982) for a discussion of this; see also the related discussion of the pairing of the arbitrary and the compulsive, 23.

17 P. Feyerabend, *Against Method* (London: Verso, 1975).

18 M. Polanyi, *The Tacit Dimension* (London: Routledge and Kegan Paul, 1967).

19 See G. Bergendahl, 'Professional skill and traditions of knowledge', in Goranzon and Florin 1990: 186; and H-G. Gadamer, *Reason in the Age of Science*, xxv, 120ff.

20 L. Code, *Rhetorical Spaces: Essays on Gendered Locations* (London: Routledge, 1995) makes this distinction throughout the essays in this book, but see 200ff.

21 L. Hunter, *Modern Allegory and Fantasy*, 1989, 169–70.

22 See R. Gooding (1989) on witnesses and lay observers to early science experiments, ' "Magnetic curves" ', 191–2.

23 See R. Gooding, as above, and S. Rose, *The Making of Memory: From Molecule to Mind* (London: Bantam Press, 1992).

24 R. Gooding, as above, xiv; see also P. Medawar, 'Is the scientific paper a fraud?', in Brown *et al.* 1986 [1967].

25 S. Shapin, 'The house of experiment', *Isis*, 79, (1988).

26 But see the more positive argument for the relation between science and technology by J. Henderson in Geiss and Viswanathan 1986: 315ff.

27 C. Bazerman, *Shaping Written Knowledge: The Genre and Activity of the Experimental Article in Science* (Madison WI: University of Wisconsin Press, 1987).
28 See the collection, *The Uses of Experiment: Studies in the Natural Sciences*, R. Gooding *et al.* (eds) (Cambridge: Cambridge University Press, 1989).
29 R. Rorty, *Objectivity, Relativism and Truth*, 8.
30 R. Rorty, as above, 10.
31 See L. Hunter, *Modern Allegory and Fantasy*, 1989.
32 R. Rorty, as above, 25.
33 See L. Hunter, *Rhetorical Stance* (1984) Part II.
34 G. Cantor, 'The rhetoric of experiment' in Gooding *et al.* 1989: 161.
35 Z. Bauman, *Modernity and Ambivalence*, 12; R. Rorty, *Objectivity, Relativism and Truth*, 44ff.; see also A. Giddens, and the discussion in Chapter 1 above.
36 R. Rorty, as above, 210.
37 See Z. Bauman on the ambivalence of tolerance, as above, 237.
38 For a helpful version of Althusser's 'Ideology and ideological state apparatuses', see D. Macdonell, *Theories of Discourse* (Oxford: Blackwell, 1986).
39 R. Rorty, as above, 36.
40 R. Rorty, as above, 39.
41 R. Williams, paper presented at the Linguistics and Literature conference, Strathclyde University, 1987.
42 T. Cave, *Recognitions: A Study in Poetics* (Oxford: Clarendon Press, 1986); see also the comments on L. Irigaray and J. Derrida in Chapter 5.
43 B. Goranzon, 'The practice of the use of computers: a paradoxical encounter between different traditions of knowledge', in Goranzon and Josefson 1990; see also D. Miall, 'An expert system approach to the interpretation of literary structure', in Ennals and Gardin 1990.
44 These are often known as theories of difference; but without social context they have no significance.
45 See B. Goranzon, as above.
46 M. Wagman, *Artificial Intelligence and Human Cognition: A Theoretical Intercomparison of Two Realms of Intellect* (London: Praeger, 1991).
47 R. Brooks, 'Intelligence with representation', *Artificial Intelligence*, 47 (1991), 142; W. Clancey, 'Model construction operators', *Artificial Intelligence*, 56 (1992), 194.
48 I. Miles *et al.* (eds) *Information Horizons: The Long-term Social Implications of New Information Technologies* (Aldershot: Edward Elgar, 1988); A. Molina, *The Social Basis of the Microelectronics Revolution* (Edinburgh: Edinburgh University Press, 1989).
49 J. Greene, *Memory, Thinking and Language* (London: Methuen, 1986).
50 H. Dreyfus, 'The Socratic and Platonic basis of cognitivism', *AI and Society*, 2: 107 (1988).
51 On not doing this, see P. Stockinger, 'Logicist analysis and conceptual inferences', and J. Fargue, 'Remarks on the interrelations between artificial intelligence, mathematical logic and humanities', both in Ennals and Gardin 1990.
52 P. Johnson-Laird, 'Human experts and expert systems', in Murray and Richardson 1989.
53 R. Ennals, *Artificial Intelligence and Human Institutions* (London: Springer-Verlag, 1991) 41.
54 R. Penrose, *The Emperor's New Mind: Concerning Computers, Minds and the Laws of Physics*, fwd by M. Gardner (London: Oxford University Press, 1989).
55 A. Janik, 'Tacit knowledge, working life and scientific "method" ', in Goranzon and Josefson 1987.
56 J. Lyotard, *The Postmodern Condition*, 1979.

57 L. Wittgenstein, *Philosophical Investigations*, trans. G. Anscombe (Oxford: Basil Blackwell, 1967 [1953]); M. Polanyi, *The Tacit Dimension*, 1967.
58 See R. Rorty, *Objectivity, Relativism and Truth*, 1991.
59 See the essays in T. Winograd and F. Flores, *Understanding Computers and Cognition: A New Foundation for Design* (Norwood NJ: Addison-Wesley, 1986); R. Ennals and J.-C. Gardin (eds) *Interpretation in the Humanities: Perspectives from Artificial Intelligence* (British Library: Library and Information Research Report 71, 1990); and R. Ennals, *Artificial Intelligence and Human Institutions*, 1991.
60 M. Cooley, *Architect or Bee?: The Human/Technology Relationship*, comp. and ed. S. Cooley (Slough: Hand and Brain Publications, 197?); and 'Creativity, skill and human-centred systems', in Goranzon and Josefson 1987.
61 R. Ennals, *Artificial Intelligence and Human Institutions*, 1991, 71.
62 M. Boden, 'Artificial intelligence: opportunities and dangers', in Murray and Richardson 1989: 166.
63 H. Dreyfus, 'The Socratic and Platonic basis of cognitivism', *AI and Society*, 2 (1988), 105.
64 For an extensive discussion, see Chapter 2 of L. Hunter, *Rhetorical Stance*, 1984.
65 A. Janik, 'Tacit knowledge, working life and scientific "method"', in Goranzon and Josefson 1987: 55.
66 This is of course the field of poetics and will be explored extensively in Chapter 6; earlier discussions of the unarticulated have focused on finding language for women's experience, H. Cixous, *'Coming to Writing' and Other Essays*, ed. D. Jenson, intro. S. Suleiman, trans. S. Cornall (Cambridge MA: Harvard University Press, 1988), and for the experience of races other than the dominant, H. Bhabha, 'Interrogating identity', *Identity Documents*, 6 (1986).
67 L. Wittgenstein, *Philosophical Investigations*, 1953, pt 159, 610.
68 P. Connerton, *How Societies Remember* (Cambridge: Cambridge University Press, 1989).
69 L. Wittgenstein, *Philosophical Investigations*, 1953, pt 179e.
70 J. Austin, *How to Do Things with Words* (Oxford: Clarendon Press, 1962).
71 For an extended discussion of late medieval to modern uses of topoi, see L. Hunter (ed.) *Topos, Commonplace, Cliché: Toward an Understanding of Analogical Reasoning* (London: Macmillan, 1991).
72 L. Hunter, 'Watson and McLuhan's *From Cliché to Archetype*', in L. Hunter (ed.) *Topos, Commonplace, Cliché*, 1991.
73 A. Janik, 'Tacit knowledge, rule-following and learning', in Goranzon and Florin 1990: 50.
74 A. Janik, 'Tacit knowledge, working life and scientific "method" ', in Goranzon and Josefson 1987: 48.
75 L. Wittgenstein, *Philosophical Investigations*, 1953, 44.
76 A. Janik, 'Tacit knowledge, working life and scientific "method" ', in Goranzon and Josefson 1987: 52–3.
77 A. Janik, 'Tacit knowledge, rule-following and learning', in Goranzon and Florin 1990: 15.
78 See the discussion at the conclusion of L. Hunter, *Modern Allegory and Fantasy*, 1989.
79 A. Janik disagrees with and quotes from H. Dreyfus, in 'Tacit knowledge, rule-following and learning', in Goranzon and Florin 1990: 48.
80 M. Polanyi, *The Tacit Dimension*, 1967, 24.
81 This link may date quite specifically from the eighteenth century, although the association needs further research. G. Cantor quotes the scientist Robert Millikan saying of an experiment, 'Beauty. *Publish* this surely, *beautiful*' ('The rhetoric of experiment', in Gooding *et al.* 1989: 159–60.

82 R. Penrose, *The Emperor's New Mind*, 1989, 26.
83 For example, see A. Sloman, 'The emperor's new mind', *Artificial Intelligence*, 56 (1992).
84 M. Serres, *Hermes: Literature, Science, Philosophy*, eds J. Harari and D. Bell, postface I. Prigogine and I. Stengers (London: Johns Hopkins University Press, 1983 [1982]), 89.
85 J. Lacan sets the stage for this movement in the essays delivered during the 1950s and 1960s, collected in *Ecrits* (1976).
86 For further commentary on this aspect of the political positioning of the individual, see Z. Bauman, *Intimations of Postmodernity* (London: Routledge, 1992) on 'competent consumers', 259.
87 J. Doran, 'Distributed artificial intelligence and the modelling of socio-cultural systems', in Murray and Richardson 1989: 82; see also R. Rorty, *Objectivity, Relativism and Truth*, 1991, 39.
88 N. Frude, 'Intelligent systems off the shelf: the high street consumer and artificial intelligence', in Murray and Richardson 1989: 162.
89 M. Polanyi, *The Tacit Dimension*, 1967, 24.
90 H.-G. Gadamer, *Reason in the Age of Science*, 1976, 90.
91 J. Habermas, *The Philosophical Discourse of Modernity*, 1987, 194.
92 S. Woolgar, 'The turn to technology in social studies of science', *Science, Technology and Human Values*, 16:1 (1991), 44.
93 See D. Gooding, ' "Magnetic curves" and the magnetic field', in Gooding *et al.* 1989.
94 R. Brooks, 'Intelligence without representation', *Artificial Intelligence*, 47 (1991) 140.
95 D. Kirsh, 'Today the earwig, tomorrow man?', *Artificial Intelligence*, 47 (1991) 161–84.
96 D. Kirsch, 'Foundations of AI: the big issues', *Artificial Intelligence*, 47 (1991) 21.
97 M. Clancey, 'Model construction operators', *Artificial Intelligence*, 56 (1992).
98 M. Clancey, 'Notes on "heuristic classification" ', *Artificial Intelligence*, 59 (1993) 193.
99 L. Birnbaum, 'Rigor mortis: a response to Nilsson's "Logic and artificial intelligence" '; D. Kirsh, 'Foundation of AI: the big issues' and 'Today the earwig, tomorrow man?'; and B. Smith, 'The owl and the electric encyclopedia', all in *Artificial Intelligence*, 47 (1991).
100 Much recent work on the early elaboration of public demonstrations of science indicates that this refusal of textuality has been present at least since the seventeenth century; see for example, S. Shapin, 'The house of experiment', *Isis*, 79 (1988).

4 The socialising of context

1 S. Turkle, *The Second Self: Computers and the Human Spirit* (London: Granada, 1984) 269.
2 Turkle 1984: 102ff.
3 S. Turkle and S. Papert, 'Epistemological pluralism: styles and voices within the computer culture', *Signs*, 16:1 (autumn, 1990).
4 The references to Aristotle's *Politics*, *Rhetoric* and *Topics* are taken from the edition J. Barnes (ed.) *The Complete Works of Aristotle, Revised Oxford Translation*, Bollingen Series LXXI vols 1 and 2 (Princeton NJ: Princeton University Press, 1984); the specific references here are to *Politics*, 1279b, and *Rhetoric*, 2173.
5 Aristotle, *Politics*, 1286b.
6 Aristotle, *Politics*, 1391ff.

7 For an extended analysis of Orwell's portrayal of this development, see L. Hunter *George Orwell: The Search for a Voice* (Open University Press, 1984).

8 Aristotle, *Politics*, 1315b.

9 This potential for the topics to alter their effect is described in L. Hunter, 'From cliché to archetype', in L. Hunter (ed.) *Topos, Commonplace and Cliché* (Macmillan, 1991); it has elements in common with J. Neel's commentary on the way that the demonstrative philosopher controls the metaphoricity of rhetoric, *Aristotle's Voice* (Carbondale and Edwardsville IL: University of Illinois Press, 1994) 178–80.

10 It is notable that a number of commentators in North America are beginning to use this interpretation of 'consensus', for example J. Poulakos in *Sophistical Rhetoric in Classical Greece* (Columbia SC: University of South Carolina Press, 1995) 150, or N. Reynolds, 'Ethos as location: new sites for understanding discursive authority', *Rhetoric Review*, 11:2 (spring, 1993). The use of 'consensus' in this way may come from the parallel use in many books on postmodernism, particularly the writing of J. Lyotard, or it may come from recent discourse theory; however, in political commentary in Europe, it still retains a usage of a communal process that can include disagreement and conflict, as in the work of Gadamer, Habermas and Beetham, whose work is discussed later in this essay.

11 L. Hunter, 'From cliché to archetype', cited above, offers a detailed analysis of this elision through M. McLuhan and W. Watson's work on the concept of 'sonambulism'.

12 Aristotle, *Politics*, 1305a.

13 For commentaries on Aristotle, women and politics, see E. Spelman, 'Aristotle and the politicization of the soul' and J. Hicks Stiehm, 'The unit of political analysis: our Aristotelain hangover', both in Harding and Hintikka 1983. For an argument that Aristotle's 'science' is not privileged but a special case de-scription, see L. Hunter, 'AI and representation: a study of a rhetorical context for legitimacy', *AI and Society*, 7 (1993) 185–207.

14 C. J. Swearingen discusses *Rhetoric* as a 'study' not an action, cited above, 118; she also distinguishes between the topics as 'signs' for the logical philosopher and as common places for rhetor. For further discussion of the philosopher as demonstrative and not social, see L. Hunter 'From cliché to archetype', cited above.

15 Aristotle, *Topics*, 155.

16 This I have argued on a number of occasions; for a summary see L. Hunter, *Rhetorical Stance in Modern Literature*, 1984, 54–7.

17 Despite the comment by J. Poulakos, cited above, that we no longer think teachers corrupt if they accept money for their work, that exchange of money does compromise them in the aims and objectives of the educational system they serve.

18 See the works cited above by S. Jarratt, C. J. Swearingen, and J. Poulakos.

19 Cited above, 15.

20 As well as Isocrates, according to the analysis by J. Poulakos, cited above.

21 One point argued by Swearingen with which I need to debate, is her claim that Plato deals with this plurality by stopping the circulation of vocabularies (99); I would suggest that he deals with plurality by evaluating and criticising the effective moral action in each vocabulary.

22 See J. Martin, *Antike Rhetorik: Technik und Methode* (Munich, 1974) 28ff., for the similar role of topos in stasis theory.

23 M. Leff, 'The topics of argumentative invention in Latin rhetorical theory from Cicero to Boethius', *Rhetorica*, 1, 1 (spring, 1983).

24 N. Streuver, 'Lorenzo Valla: humanist rhetoric and the critique of the classical languages of morality', in J. J. Murphy (ed.) *Renaissance Eloquence: Studies in the Theory and Practice of Renaissance Rhetoric* (London, 1983).

25 C. Perelman and L. Olbrechts-Tyteca, *The New Rhetoric: A Treatise on Argumentation*, trans. J. Wilkinson and P. Weaver (London, 1971).

26 L. Hunter, 'A rhetoric of mass communication: collective or corporate public discourse', in R. Enos (ed.) *Oral and Written Communication: Historical Approaches* (London: Sage, 1990).

27 B. Stock, *The Implications of Literacy: Written Language and Models of Interpretaion in the Eleventh and Twelfth Centuries* (Princeton NJ: Princeton University Press, 1983); and M. Spufford, *Small Books and Pleasant Histories* (Cambridge: Cambridge University Press, 1981) 8.

28 For one pertinent study exploring this issue, see L. Jardine, *Erasmus, Man of Letters* (London: Cambridge University Press, 1994).

29 For example, see A. Giddens 'Structuralism, Post-structuralism and the Production of Culture' in A. Giddens and J. Turner (eds) *Social Theory Today* (Cambridge: Polity Press, 1987) 202.

30 See L. Hunter, *Rhetorical Stance*, 1984.

31 E. Havelock, *Preface to Plato* (Cambridge MA: Harvard University Press, 1963)

32 Plato, *The Republic*, trans. D. Lee (Harmondsworth: Penguin, 1973) 93.

33 For a detailed analysis, see L. Hunter, *Rhetorical Stance*, 1984, Chapter 2.

34 Plato, *Phaedrus*, trans. W. Hamilton (Harmondsworth: Penguin, 1973 [1955]); all references are taken from this edition.

35 J. Derrida, *Of Grammatology*, trans. G. Spivak (London: Johns Hopkins University Press, 1974).

36 B. Stock, *The Implications of Literacy*, 1983, 19.

37 B. Pattison, *On Literacy: The Politics of the Word from Homer to the Age of Rock* (Oxford: Oxford University Press, 1982) 99. See also L. Hunter, 'A rhetoric of mass communication', 1990.

38 L. Hunter, 'A rhetoric of mass communication', 1990, 234–6.

39 D. Olson and A. Hildyard, 'Writing and literal meaning', in M. Martlew (ed.) *The Psychology of Written Languages: Developmental and Educational Perspectives* (Chichester: Wiley, 1983).

40 S. Scribner and M. Cole, *The Psychology of Literature* (London: Harvard University Press, 1981).

41 P. Tulviste, 'On the origins of theoristic syllogistic reasoning in culture and the child', *Quarterly Newsletter of Comprehensive Human Cognition*, 1 (1979).

42 J. J. Murphy (ed.) *The Rhetorical Tradition and Modern Writing* (New York: Modern Languages Association, 1982).

43 T. Sloan, 'The crossing of rhetoric and poetry in the English renaissance', in T. Sloan and C. Waddington (eds) *The Rhetoric of Renaissance Poetry from Wyatt to Milton* (London: University of California Press, 1974).

44 On Whateley and Blair see N. Johnson, 'Three nineteenth-century rhetoricians', in J. J. Murphy (ed.) *The Rhetorical Tradition and Modern Writing* (New York: Modern Languages Association, 1982); for the research on later periods see J. Cook-Gumperz (ed.) *The Social Construction of Literacy* (London: Cambridge University Press, 1986); P. Tulviste, 'On the origins of theoristic syllogistic reasoning in culture and the child', *Quarterly Newsletter of Comprehensive Human Cognition*, 1 (1979); and J. Vgotsky, 'The prehistory of written language', in M. Martlew, *The Psychology of Written Language, Developmental and Educational Perspectives* (Chichester: Wiley, 1983).

45 I am grateful to Lesley Johnson for her comments on this field. See M. Parke, 'Influence of concepts of *Ordinatio* and *compilatio* in the development of the book', in J. Alexander and M. Gibson (eds) *Medieval Learning* (Oxford, 1976);

and B. Stock, as above (1983); and J. M. Lechner, *Renaissance Concepts of the Commonplaces* (Westport RI, 1962) 679. Lechner also quotes F. Bacon on the topic as a 'place where a thing is to be looked for, marked and indexed' (68). See also F. Goyet, 'The word "commonplaces" in Montaigne', in L. Hunter (ed.) *Toward a Definition of Topos: Approaches to Analogical Reasoning* (London: Macmillan, 1991).

46 A. Grafton and L. Jardine, *From Humanism to the Humanities: Education and the Liberal Arts in Fifteenth- and Sixteenth-Century Europe* (London: Duckworth, 1986).

47 F. Bacon, *The Advancement of Learning*, in *Works of Bacon, VI*, 296; and W. Howell, *Logic and Rhetoric in England, 1500–1700* (Princeton NJ, 1956), 353–4.

48 W. Howell, as above, 353–4.

49 See L. Hunter, 'Watson and McLuhan's *From Cliché to Archetype*' (1991) for an account of one part of this activity in the field of criticism and theory in literature.

50 See Ong 1971; L. Hunter 1984; and more recently S. Woolgar, 'The ideology of presentation and the role of the agent', in Lawson and Appignanesi 1989.

51 H.-G. Gadamer, *Reason in the Age of Science*, trans. F. Lawrence (London: MIT Press, 1981 [1976]) 77.

52 J. Habermas, *The Philosophical Discourse of Modernity* (Polity, 1987) 194.

53 Not surprisingly, medicine which has a focus on the necessary context of the human body, is also the site where science and ethics are most under debate. Medicine is also the site for the most consistent consideration of the social implications of persuasive strategies in rhetoric, beginning as N. Streuver notes with Descartes in 'The discourse of cure; rhetoric and medicine in the late Renaissance', given as the presidential address to the International Society for the History of Rhetoric, Gottingen, 1989.

54 Artificial intelligence research has, since the 1960s, consisted of attempts to restore context to scientific methodology largely through formal systems such as frames or connectionist neural networks; see the debate around contributors recounted in Chapter 2 of Graubard 1988.

55 Such an awareness is indicated not only in S. Hockey's early work, *Guide to Computer Applications in the Humanities* (London: Duckworth, 1980) which stresses computer-aided research, but also in a number of articles concerning computer-based learning.

56 G. Landow, *Hypertext: The Convergence of Contemporary Critical Theory and Technology* (London: Johns Hopkins University Press, 1992).

57 No mention is made, for example, of either the *Talmud* or of concrete poetry.

58 An area where a similar worry about the reader's access to forms of the text has already emerged, is in the discussion among textual editors about the suitability of providing parallel texts of different editions of a single book. In computer-readable form these parallel texts permit the reader immense flexibility, and many editors are concerned about the implications.

59 H. Mandel and J. Levin (eds) *Knowledge Acquisition from Texts and Pictures* (Amsterdam: Elsevier Science Publications, 1988).

60 W. Beeman *et al.*, 'Hypertext and pluralism: from lineal to non-lineal thinking', in *Proceedings of Hypertext '87*.

61 Handout for XANADU presented to the Dynamic Text Conference '89.

62 A helpful warning on this lack of education comes from M. Elsom-Cook, in 'Multimedia: the emperor's new clothes', the editorial for *ALTNEWS, New Technologies and Learning in Europe*, 6 (February 1991).

63 Hypermedia users often claim that the texts can be 'unstructured'. D. Jonassen, in *Hypertext/Hypermedia*, 51, says 'Unstructured hypertext is random, node-link hypertext'. But selection of texts (nodes) is structure in itself.

64 Many of the papers from the Hypertext '87 conference were concerned with this aspect; for example see D. Charney, 'Comprehending non-linear text: the role of discourse cues and reading strategies', in *Proceedings of Hypertext '87*.

65 Problems of 'navigation' are a central focus for hypermedia researchers. See for example, L. Hardman and D. Edwards, ' "Lost in hyperspace": cognitive mapping and navigation in a hypertext environment', in McAleese 1989; or R. Wright and A. Lickorish, 'The Influence of discourse structure on display and navigation in hypertexts', in Williams and Holt 1989.

66 HYPERDOC was used as a display medium for introductory work on the Hartlib Papers Project.

67 Boyce, G. *et al.* (eds) *Newspaper History from the Seventeenth Century to the Present Day* (London: Constable, 1978).

68 R. Jones, 'Hypertext-based, integrated laboratory information system', Text Retrieval conference, 1989.

69 See F. Goyet, 'Common places in Montaigne', in L. Hunter (ed.) (1991), and L. Hunter, 'Watson and McLuhan's *From Cliché to Archetype*' in the same collection.

70 An account of the project may be found in L. Hunter *et al.*, 'Creating a hypertext with guide', in J. Darby (ed.) *The CTISS File*, 9 (February 1990).

71 M. Leslie, 'The Hartlib Papers Project: text retrieval with large datasets', *JALLC*, 5:1 (1990).

72 Dr R. Jones noted an analogous procedure in science in the recognition of insulin periodicity, 'Hypertext-based, integrated laboratory information system', Text Retrieval conference, 1989.

73 See for example the work by J. Shattock *et al.* on the Victorian Periodicals Project; and the studies offered by the *Victorian Periodicals Newsletter*.

74 M. Beetham, *A Magazine of her Own: Domesticity and Desire in the Woman's Magazine, 1800–1914* (London: Routledge, 1996); see also M. Angenot, 'Social discourse analysis: outlines of a research project', *Discours Social/Social Discourse*, 1, 1 (1988).

75 N. Norman, 'The electronic teaching theatre: interactive hypermedia and mental models of the classroom', in Ward 1990.

76 L. West and A. Pines (eds) *Cognitive Structure and Conceptual Change* (London: Academic Press, 1985).

77 The implications of the computer simply as a repository for data is discussed in *Cognitive Science and its Applications for Human Computer Interaction* (London: Lawrence Erlbaum Associates, 1988).

5 Feminist critiques of science

1 For an elaboration of this argument see H. Arendt, *The Origins of Totalitarianism* (New York: Harcourt, Brace, Jovanovitch, 1951).

2 D. Beetham and K. Boyle in *Introducing Democracy: 80 Questions and Answers* (Polity/UNESCO, 1995), offer a remarkably clear and helpful guide to contemporary democratic structures deriving from Eurocentric political philosophy.

3 See A. Giddens 'Structuralism, post-structuralism and the production of culture', in A. Giddens and J. Turner (eds) *Social Theory Today* (Cambridge: Polity, 1987) on the lack of agency in structuralism and its derivatory criticisms.

4 See for example the extensive discussion concerning the way ahistorical postmodernism empties out agency, between S. Harding and C. DiStefano in *Whose Science? Whose Knowledge? Thinking from Women's Lives* (Milton Keynes: Open University Press, 1991) 182ff.

5 For just one distinction between the two postmodernisms, see A. Yeatman, 'Postmodern epistemological politics and social science', in K. Lennon and M.

Whitford (eds) *Knowing the Difference: Feminist Perspectives in Epistemology* (London: Routledge, 1994) 187.

6 S. Hekman, *Gender and Knowledge: Elements of a Postmodern Feminism* (Polity/Blackwell, 1990).

7 The publication of B. Latour and S. Woolgar, *Laboratory Life: The Social Construction of Scientific Facts* (Beverley Hills CA: Sage, 1979) has moved in just fifteen years from being a text central to the critique of science to being a text central to a critique of relativism which it seems to support; but see B. Latour, 'Clothing the naked truth', in H. Lawson and L. Appingnanesi (eds) *Dismantling Truth* (London: Wiedenfeld and Nicholoson, 1989).

8 This has been proposed elsewhere in my work; see L. Hunter, *Rhetorical Stance in Modern Literature*, 1984.

9 T. de Lauretis, 'Eccentric subjects: feminist theory and historical conscious-ness', *Feminist Studies*, 19:1 (spring, 1990) 123.

10 H. Longino, 'Subjects, power and knowledge: description and prescription in feminist philosophies of science', in E. Fox-Keller and H. Longino (eds) *Femi-nism and Science* (Oxford: Oxford University Press, 1990) 268; reprinted from L. Alcoff and E. Potter (eds) *Feminist Epistemologies* (New York: Routledge, 1993).

11 D. Smith, *The Everyday World as Problematic: A Feminist Sociology* (Boston: Northeastern University Press, 1987).

12 See particularly 119–37.

13 Cited above.

14 L. Code, *Rhetorical Spaces: Essays on (Gendered) Locations* (London: Routledge, 1995). Code's use of the word 'rhetoric' points to the inexorably social materiality of gender, but she does not bring to her analysis any of the work in the history of rhetoric on textuality and stance. Hence her work tends to focus on ethos.

15 The baseline text in this area is D. Winnicott, *Playing and Reality* (New York: Basic Books, 1971) and is followed by N. Chodorow, *The Reproduction of Mothering, Psychoanalysis and the Sociology of Gender* (Berkeley CA: Univer-sity of California Press, 1978); and D. Dinnerstein, *The Mermaid and the Mino-taur: Sexual Arrangements and Human Malaise* (New York: Harper and Row, 1976). Recent feminist theorists using standpoint include E. Fox-Keller, see for example 'Feminism and science', *Signs*, 7/3 (1982), S. James 'The good-enough citizen: female citizenship and independence', in Bock and James 1992; and much of J. Flax's work, for an early example see 'Political philosophy and the patriarchal unconscious', in Harding and Hintikka 1983.

16 Central to this field is C. Gilligan, *In a Different Voice: Psychological Theory and Women's Development* (Cambridge: Harvard University Press, 1982); her work is broadened and extended by for example, M. Field Belenky *et al.*, *Women's Ways of Knowing: The Development of Self, Voice and Mind* (New York: Basic Books, 1986).

17 Classics in this field are G. Anzaldua and C. Moraga (eds) *This Bridge Called My Back: Writings by Radical Women of Colour* (Watertown MA: Persephone Press, 1981), and P. Hill Collins, *Black Feminist Thought: Knowledge, Con-sciousness and the Politics of Empowerment* (London: Harper Collins Aca-demic, 1990).

18 See for example, I. Diamond and L. Quinby (eds) *Feminism and Foucault: Reflections on Resistance* (Boston MA: Northeastern University Press, 1988); and C. Ramzouoglu (ed.) *Up Against Foucault: Explorations of some Tensions Between Foucault and Feminism* (London: Routledge, 1993).

19 D. Smith, *The Everyday World as Problematic*, 3.

20 'Beyond equality: gender, justice and difference', in Bock and James 1992: 203.

21 A. Jaggar, *Feminist Politics and Human Nature* (Brighton: Harvester, 1983) 307.

22 See for example the introduction to K. Lennon and M. Whitford, *Knowing the Difference* (London: Routledge, 1994) 2–4, where a critique of standpoint as perspectival leads them to advocate historical postmodernism.

23 H. Rose, *Love, Power and Knowledge: Towards a Feminist Transformation of the Sciences* (Cambridge: Polity, 1994).

24 *Love, Power and Knowledge*, 208ff.

25 S. Harding in *The Science Question in Feminism* (Milton Keynes: Open University Press, 1986) 245; she also takes up the importance of metaphor (233) and expands on this in *Whose Science?* to look at literary context and the non-linearity of textual devices as democratic and womanly (301).

26 D. Haraway, 'A manifesto for cyborgs: science, technology and socialist feminism in the 1980s', in L. Nicholson (ed.) *Feminism/Postmodernism* (London: Routledge, 1990 [1985]); and D. Haraway, 'Situated knowledges: the science question in feminism and the privilege of partial perspective', *Feminist Studies*, 14/3 (1988), 575–99.

27 E. Fox-Keller, *Secrets of Life, Secrets of Death: Essays on Language, Gender and Science* (London: Routledge, 1992) throughout Chapters 6 to 9.

28 L. Code, *Rhetorical Spaces*, particulary in 'Voice and voicelessness', 154ff.

29 For example, see C. Merchant *The Death of Nature: Women, Ecology and the Scientific Revolution* (London: Wildwood House, 1980); Linda Schiebinger *The Mind has no Sex? Women in the Origins of Modern Science* (Cambridge MA: Harvard University Press, 1989); and Genevieve Lloyd, *The Man of Reason: 'Male' and 'Female' in Western Philosophy* (Minneapolis MN: University of Minnesota Press, 1984).

30 See L. Hunter and S. Hutton (eds) *Women in Medicine, Science and Technology, 1500–1700* (Stroud: Alan Sutton, 1997).

31 In a study that precisely engages with problematising assumptions, S. Harding bases one revision on this claim using the work of E. Zilsel, *The Science Question*, 218–20. Similar claims to literacy levels are made by many, for example S. Hekman, who in an attack on Bacon in *Gender and Knowledge*, dismisses early women as scientists because they could not speak Latin.

32 See W. Eamon, *Science and the Secrets of Nature: Books of Secrets in Medieval and Early Modern Culture* (Princeton NJ: Princeton University Press, 1994); see also L. Hunter, 'Household Books 1500–1700', in *Cambridge Bibliography to English Literature, 1500–1700*, Third Edition (Cambridge, 1997).

33 S. Shapin, 'The house of experiment in seventeenth-century England', *Isis*, 79 (1988).

34 See W. Wallace 'Aristotelian science and rhetoric in transition: the middle ages and the Renaissance' and J. Dietz Moss, 'The interplay of science and rhetoric in seventeenth century Italy', both in *Rhetorica*, VII:1 (winter, 1989) for helpful and detailed discussion of the movement from medieval to Renaissance to post-Renaissance science.

35 L. Hunter, 'Women and domestic medicine', in Hunter and Hutton 1997.

36 There is a consistent history of publication in the name of the poor who cannot afford professional help from *This is the Glasse of Helthe* (1540), to John Partridge's *A Treasurie of Commodious Conceits and Hidden Secrets* (1573?), to writers such as Christopher Bennett and his *Health's Improvement* (1655) and Nicholas Culpeper, whose translation of the College of Physicians' *Pharmacopoeia* was undertaken in order to make the remedies more generally available. See P. Slack, 'Mirrors of health and treasures of poor men', in C. Webster (ed.) *Health Medicine and Mortality in the Sixteenth Century* (Cambridge: Cambridge University Press, 1979).

37 There is extensive evidence from diaries and letters, and from the documents of apprenticeships and so on, that women undertook many of the surgical and herbal responsibilities for their communities, as Margaret Pelling has recon-

structed in for example 'Apprenticeship, health and social cohesion in early modern London', *History Workshop Journal*, 37 (1994). Aristocratic ladies of the sixteenth century frequently took over directing activities that would have fallen to the monasteries before 1533. Holinshed notes this area, quoting W. Harrison in 1586, *Early English Meals and Manners*, ed. F. J. Furnivall (Early English Text Society, OS32, 1868) xc.

38 John Moore writes to Samuel Hartlib in the early 1650s, having asked Hartlib how he might find paying employment, that the suggestion that he get some of Benjamin Worsley's secret receipts and prepare them commercially is a non-starter since Worsley of all people keeps his secrets to himself.

39 R. Maddison in *The Life of the Hon. Robert Boyle* (London: Taylor and Francis, 1969) suggests that the commercial implications were the reason that Boyle's early scientific circle called themselves the 'Invisible College'.

40 See S. Shapin, ' "A scholar and a gentleman": the problematic identity of the scientific practitioner in early modern England', *Journal of the History of Science*, XXIX (1991); see also A. Johns, 'History, science, and the history of the book: the making of natural philosophy in early modern England', *Publishing History*, 30 (1991) 8.

41 L. Hunter, 'Sisters of the Royal Society', in Hunter and Hutton 1997.

42 For one example among many see M. diLeonardo, 'Contingencies of value in feminist anthropology' in J. Hartman and E. Messer-Davidow (eds), *(En)Gendering Knowledge: Feminists in Academe* (Knoxville: University of Tennessee Press, 1991) 150.

43 See N. Streuver, 'The discourse of cure: rhetoric and medicine in the late Renaissance', Presidential Paper given to the International Society for the History of Rhetoric, Göttingen 1989.

44 This work was probably written in the late 1640s although printed by Hartlib in 1655, and J. T. Harwood in *Robert Boyle: The Early Essays and Ethics of Robert Boyle* (Carbondale IL: Southern Illinois University Press, 1991) suggests that Boyle may have got this idea from his sister Katherine. It was certainly common for women to attribute receipts to their sources.

45 John Aubrey, in *Brief Lives*, ed. A. Clark (Oxford: Clarendon Press, 1989) says that Mary Sidney had a laboratory used by her brother Philip and her husband William Cavendish.

46 The most well documented communicative space is that that of the Hartlib circle, and there is growing evidence of circles of aristocratic women sharing diary and letter extracts not only in science but in the arts.

47 Hekman 1990: 127.

48 R. Doell, 'Whose research is this? Values and biology', in Hartman and Messer-Davidow, 122.

49 See for example L. Jordanova, *Sexual Visions: Images of Gender in Science and Medicine between the Eighteenth and Twentieth Centuries* (Brighton: Harvester Wheatsheaf, 1989).

50 K. Digby (trans.) *A Treatise of Adhering to God* (London: printed for Henry Herringman, 1654) A4v.

51 E. Fox-Keller, *Secrets of Life, Secrets of Death*, 28.

52 N. Scheman, 'Who wants to know? The epistemological value of values', in Hartman and Messer-Davidow, 179.

53 See L. Code, 'Taking subjectivity into account', in *Rhetorical Spaces*, 23ff.

54 S. Lovibond, 'The end of morality?', in Lennon and Whitford, 71.

55 J. Elshtain, 'The power and powerlesness of women', in Bock and James, 113.

56 N. Fraser, 'What's critical about critical theory? The case of Habermas and gender', in M. Shanley and C. Pateman (eds) *Feminist Interpretation and Political Theory* (Cambridge: Polity, 1991) 257.

57 N. Fraser, *Unruly Practices: Power, Discourse and Gender in Contemporary Social Theory* (Cambridge: Polity, 1989) 117.

58 H. Rose, *Love, Power and Knowledge*, 25.

59 J. Sawicki, 'Foucault and feminism: toward a politics of difference', in Shanley and Pateman, 221.

60 L. Code also notes this in *Rhetorical Spaces*, 104.

61 See particularly S. Harding in *Whose Science?*, and L. Code in *Rhetorical Spaces*, who both from early articles by H. Rose such as 'Hand, brain and heart: toward a feminist epistemology for the natural sciences', *Signs*, 9 (1, 1983) 73–96.

62 N. Scheman, 'Who wants to know?', in Hartman and Messer-Davidow, 179.

63 See the introduction to Diamond and Quinby 1988.

64 L. Code, *Rhetorical Spaces*, 106.

65 D. Haraway, 'Situated knowledges: the science question in feminism and the privilege of partial perspective', *Feminist Studies*, 14/3 (1988) 575–99.

66 Code derives this notion from L. Wittgenstein, particulary *The Blue and Brown Books* that I personally have always found useful for negotiating between the flexible and the relativist or the rhetorically probable and plausible, and which are increasingly used by feminist philosophers such as A. Tanesini and S. Lovibond.

67 D. Haraway, 'Situated knowledges', 583.

68 M. Fricker, 'Knowledge as construct: theorizing the role of gender in knowledge', in Lennon and Whitford, 98.

69 S. Lovibond, 'The end of morality?', in Lennon and Whitford, n71.

70 A. Tanesini, 'Whose language?', in Lennon and Whitford, 108.

71 H. Rose, *Love, Power and Knowledge*, 74, 93.

72 See S. Harding, *The Science Question*, 139.

73 For one example among many see L. Code, *Rhetorical Spaces*, 53.

74 C. di Stephano, 'Masculine Marx', in Shanley and Pateman, 152.

75 S. Harding, see especially the chapter on 'Strong objectivity' in *Whose Science?*, 138ff.

76 L. Code, 'Who cares?', in *Rhetorical Spaces*, 116 and elsewhere.

77 K. Ferguson, *The Man Question: Visions of Subjectivity in Feminist Theory* (Oxford: University of California Press, 1993) 5.

78 N. Fraser, *Unruly Practices*, 19.

79 P. Violi, 'Gender, subjectivity and language', Bock and James, 168.

80 R. Braidotti, 'On the female subject, or: from "she-self" to "she-other" ', in Bock and James, 187; Irigaray's work on mimesis is congruent with that of J. Derrida.

81 See also L. Hunter, *Modern Allegory and Fantasy*, 135.

82 N. Fraser, *Unruly Practices*, 173.

83 A. Jaggar, 'Love and knowledge: emotion in feminist epistemology', in A. Jaggar and S. Bordo (eds) *Gender/Body/Knowledge: Feminist Deconstructions of Being and Knowing* (New Brunswick NJ: Rutgers University Press, 1989), 160–3.

84 H. Longino, 'Subjects, power and knowledge', in Fox-Keller and Longino, 273.

85 E. Fox-Keller, *Secrets of Life, Secrets of Death*, 94.

86 L. Code, 'Incredulity, experientialism, and the politics of knowledge', in *Rhetorical Spaces*, 58ff.

87 L. Code, *Rhetorical Spaces*, 108–9.

88 N. Fraser, *Unruly Practices*, 128–31.

89 N. Fraser, *Unruly Practices*, 151–3.

90 C. Pateman, *The Problem of Political Obligation*, 167.

91 C. Pateman, *Democracy, Freedom and Special Rights* (Swansea: University of Wales Press, 1995) 6–18.

92 J. Flax, 'Beyond equality', in Bock and James, 194.
93 These critiques are extensive, but to name a few that focus on the contest: L. Jardine, *Erasmus, Man of Letters* (Cambridge: Cambridge University Press, 1994); G. Kennedy, *Classical Rhetoric and its Christian and Secular Tradition from Ancient to Modern Times* (Chapel Hill NC: University of North Carolina Press, 1980); J. J. Murphy, *Rhetoric in the Middle Ages: A History of Rhetorical Theory from Saint Augustine to the Renaissance* (Berkeley CA: University of California Press, 1974); B. Stock, *The Implications of Literacy: Written Language and Models of Interpretation in the Eleventh and Twelfth Centuries* (Princeton NJ: Princeton University Press, 1983); N. Streuver, 'Lorenzo Valla: Humanist Rhetoric and the Critique of the Classical Language of Morality', in J. J. Murphy (ed.) *The Rhetorical Tradition and Modern Writing* (New York: Modern Languages Association); C. J. Swearingen, *Rhetoric and Irony: Western Literacy and Western Lies* (Oxford: Oxford University Press, 1991).
94 J. Flax, 'Beyond equality', in Bock and James, 193.
95 H. Rose (1997a) 'Goodbye truth, hello trust', from typescript published in *Science and the Construction of Women*.
96 P. Hill Collins, *Black Feminist Thought*, 69, 225. Hill Collins also discusses knowledge claims under the terms 'truth' and 'trust' in this work, see 202ff.
97 H. Rose, 'Rhetoric, feminism and scientific knowledge: or, from either/or to both/and', in R. Roberts and J. Goode (eds) *The Rediscovery of Rhetoric* (London: Bristol Classical Press, 1993).
98 In *The Postmodern Condition*, J. Lyotard discusses this loss of control under the pluralism of the term 'paralogy'; Rose is more concrete: her description of the development of the genome project in *Love, Power and Knowledge* is a detailed example of how the huge institutional corporate structure of technoscience can acquire an equilibrium that is self-perpetuating.
99 For a flexible corporatism, see H. Rose on Sweden, *Love, Power and Knowledge*, 66–7, and other comments on this country throughout the analysis.
100 G. Spivak, in an opening address to the conference on Postcolonial and Commonwealth Literature in Oviedo, 1996, raised these devices as particularly helpful at the moment. Having done so, it became apparent that many other conference participants were focusing on them as well. What one makes of this requires some study and thought, but the congruency underlines the way that studies in the history of rhetoric have pointed out, that the same devices become historically appropriate to different communities at the same time.
101 H. Arendt's distinction between living together and ruling government is discussed in M. Dietz, 'Hannah Arendt and feminist politics', in Shanley and Pateman; see also M. Canovan, *Hannah Arendt: A Reinterpretation of Her Political Thought* (Cambridge: Cambridge University Press, 1992).
102 Most feminist critics say that position should be indicated as far as possible, but few go on to the next step, which is actively to consider the implications of the position. Those who do include S. Jarratt, 'The first sophists and feminism: discourses of the "Other" ', *Hypatia*, 5:1 (1990) 27–41; and S. Harding in for example *Whose Science?*.
103 The history of the editing of Shakespeare is a good place to become familiar with the self-display of textuality. Just one example may indicate what I mean: Pope's edition of Shakespeare's plays, which appeared in 1723, is heavily edited. Pope changed the words on the basis that if Shakespeare had lived for another century he would have wanted to use the more elegant vocabulary of England's neoclassical movement. Pope also marked the texts throughout with special symbols indicating 'irony', 'passages of special beauty', and so on; in effect this was an early version of literary criticism as it became in the the nineteenth century, which in an attempt to justify its scientificity assumed broad authority

over the text. For one account see G. Taylor, *Reinventing Shakepeare: A Cultural History from the Restoration to the Present* (London: Hogarth, 1989).

104 See E. Fox-Genovese, *Feminism is not the Story of My Life: How Today's Feminist Elite has Lost Touch with the Real Concerns of Women* (New York: Nan A. Talese, 1996); this is difficult book to read, but raises important issues such as this in its odd conflation of messiness and complexity.

105 S. Turkle and S. Papert, 'Epistemological pluralism: styles and voices within the computer culture', *Signs*, 16:1 (1990) 136; see also the ground-breaking S. Turkle, *The Second Self: Computers and the Human Spirit* (London: Granada, 1984).

106 See R. Perry's introduction to *Signs*, 16:1 (1990) on computing, 'The computer cluster'; see also *Women into Computing: Selected Papers 1988–90* (London: Springer-Verlag, 1991) and *Women into Computing: Teaching Computing, Content and Methods* (Keele University, Staffordshire, 1992).

107 D. Haraway, 'Situated knowledges', 576.

108 See in particular *Women into Computing: Teaching Computing, Content and Methods*.

109 G. Spivak, 'Reflections on cultural studies in the post-colonial conjuncture', *Critical Studies: Cultural Studies Crossing Borders*, 3:1 (1991) 65.

6 A feminist critique of the rhetorical stance of contemporary aesthetics

1 L. Hunter, *Rhetorical Stance in Modern Literature* (London: Macmillan, 1984).

2 See P. Hill Collins, *Black Feminist Thought*, 149–52, for analysis of the ways in which education can be central to encouraging people to articulate standpoint and change social practices.

3 L. Code, 'Voice and voicelessness', in *Rhetorical Spaces: Essays on (Gendered) Locations* (London: Routledge, 1995) 184.

4 See especially 160ff.

5 J. Wolff, *Feminine Sentences: Essays on Women and Culture* (Cambridge: Polity, 1990).

6 Contributions could include Liz Stanley's work over the past decade, see her 'The knowing because experiencing subject: narratives, lives and autobiography', in K. Lennon and M. Whitford, *Knowing the Difference* (London: Routledge, 1994) 132–48; and it may be significant that Stanley is a sociologist rather than a literary critic by training, since she never explicitly takes on the aesthetic establishment. Stanley asks in 'Feminist auto/biography and feminist epistemology', in J. Aaron and S. Walby (eds) *Out of the Margins: Women's Studies in the Nineties* (London: The Falmer Press, 1991) when feminist critics will begin to write in the way that they value in literature, and Jeanne Perreault's *Writing Selves: Contemporary Feminist Autography* (Minneapolis: University of Minnesota Press, 1995) combines discussions of Audre Lorde, Adrienne Rich, Kate Millett and Patricia Williams with her own ability to write herself into the text.

7 For an implicit critique also working through biography, see L. Hunter, *George Orwell: The Search for a Voice* (Milton Keynes: Open University Press, 1984); see also the central essays in L. Hunter, *Outsider Notes* (Vancouver: Talonbooks, 1996).

8 J. Perreault, *Writing Selves*, 4.

9 P. Hill Collins, *Black Feminist Thought*, 227.

10 An analysis and summary of these critical issues is presented in L. Hunter, *Modern Allegory and Fantasy* (London: Macmillan, 1989). See also C. Brooke-Rose, *A Rhetoric of the Unreal: Studies in Narrative and Structure, Especially of*

the Fantastic (Cambridge: Cambridge University Press, 1981); and R. Jackson, *Fantasy, the Literature of Subversion* (London: Methuen, 1981).

11 See T. Todorov, *Introduction à la Littérature Fantastique* (Paris: Editions du Seuil, 1970)

12 This is the implicit import of F. Jameson's *The Political Unconscious* (London: Cornell University Press, 1981).

13 A discussion of the counter-movement, and the difficulty I have had with it, can be found in L. Hunter, *Outsider Notes: Feminist Approaches to the Nation State, Writers/Readers and Publishing* (Vancouver: Talonbooks, 1996).

14 H. Rose and S. Rose, *The Radicalisation of Science: Ideology of/in the Natural Sciences* (London: Macmillan, 1976) xvi–xvii.

15 Feminist philosophers have at times turned to the writing of H. Arendt to discuss these issues, and her vocabulary set is congruent with many writers. M. Dietz, 'Hannah Arendt and feminist politics' in Shanley and Pateman (1991: 236) indicates why when she notes Arendt's concern with the loss of 'collective capacity for exercising power through shared word and deed'. She also elaborates on the unease Arendt shows in *The Human Condition* (1958) for the growing realisation that the vote does not equal participation in democracy, which is a central element in more recent critiques of liberal democracy by C. Pateman and N. Fraser.

16 Marilyn Butler's 'Repossessing the past: the case for an open literary history', in M. Levinson *et al.* (eds) *Rethinking Historicism: Critical Readings in Romantic History* (Oxford: Blackwell, 1989) remains one of the most lucid treatments of this area. See also T. Eagleton, 'The rise of English Studies', in *Literary Theory: An Introduction* (Oxford: Blackwell, 1983); and F. Court, *Institutionalising English Literature: The Culture and Politics of Literary Study 1750–1990* (Stanford CA: Stanford University Press, 1992).

17 See C. J. Swearingen, *Rhetoric and Irony*, for an extensive discussion of these questions in the early classical period.

18 P. Hill Collins notes the variable characteristics of silence in her comments on 'rearticulation' as the movement from silence to language to action, *Black Feminist Thought*, 112.

19 I was reminded of this by Julia Emberley, whom I thank for the detail of the variety of reproduction involved.

20 For example, see G. Lloyd, *The Man of Reason*, 11, who goes on to speak of the metaphorical weight of Bacon's vocabulary as part of the shift from viewing nature as an organism to nature as a machine. However, one can alternatively think of the shift as one from nature as authority to nature as chaos.

21 I would like to thank Lesley Johnson for her generous discussions of medieval literature, feminism and women, from which this comment derives.

22 J. Perreault cites Adrienne Rich's comments on the way such amnesia creates a silence in which one could be said to 'forget one's self', *Writing Selves*, 42.

23 R. Barthes is a good example of the way that detailed critique circles back into the canon; Foucault resists the movement by increasingly locating his criticism in isolated nodes of discourse that defy connection with the daily lives of people; while Derrida tortuously performs the difficulty itself.

24 M. Laurence, *The Diviners* (New York: McClelland and Stewart/Seal, 1975 [1974]).

25 We discussed short stories mainly from A. Munro's *The Dance of the Happy Shades* (Toronto: McClelland and Stewart, 1968) and *The Progress of Love* (London: Chatto and Windus, 1985).

26 The courses looked at M. Atwood, *Cat's Eye* (London: Virago, 1990 [1988]) and *The Robber Bride* (London: Virago, 1994 [1993]).

27 For an account of the debate see L. Hunter, 'Bodily functions in Cartesian space', in S. Chew and L. Hunter (eds) *Borderblur: Poetry and Poetics in*

Canadian Literature (Edinburgh: Quadriga, 1996) 150–73, initially presented as a paper at the Borderblur conference, Leeds 1991.

28 N. Brossard, *These Our Mothers*, trans. B. Godard (Toronto: Coach House, 1983 [1977]), and *Mauve Desert*, trans. S. de Lotbinière-Harwood (Toronto: Coach House, 1990 [1987]).

29 D. Marlatt, *Steveston* (Vancouver: Talonbooks, 1974), and *Salvage* (Red Deer: Red Deer College Press, 1991).

30 J. Armstrong, *Slash* (Penticton: Theytus, 1985); B. Culleton, *In Search of April Raintree* (Winnipeg: Pemmican, 1987); E. Danica, *Don't, a Woman's Word* (Toronto: McClelland and Stewart, 1990 [1988]); R. Doiron, *My Name is Rose* (Toronto: East End Literacy Press, 1987); and J. Dumas, *Madeleine and the Angel* (Saskatoon: Fifth House, 1989).

31 See my attempts to learn how to read in *Outsider Notes*, Chapter 6.

32 Most of my British students will not discuss religion in any contemporary texts, although they seem happy to do so with earlier ones.

33 P. Hill Collins, *Black Feminist Thought*, 190.

34 J. Butler, *Bodies that Matter* (London: Routledge, 1993) develops the idea of the abject extensively in the opening chapter of that book, and glosses the development of the term from J. Kristeva and others.

35 S. Lovibond, 'The end of morality?' in Lennon and Whitford, 69.

36 S. Benhabib, 'The Generalized and the concrete other', in E. Fraser *et al.* (eds) *Ethics: A Feminist Reader* (Oxford: Blackwell, 1992) 268.

37 J. Flax, 'Beyond equality', in Bock and James, 204; see also the distinction between a unitary self and a core self in *Thinking Fragments: Psychoanalysis, Feminism, and Postmodernism in the Contemporary West* (Berkeley CA: University of California Press, 1990) 218ff.

38 S. Lovibond, 'The end of morality?' in Lennon and Whitford, 70.

39 This has resonances with M. Dietz's reading of Arendt's 'politics of difference', in which to act, humans must have a 'private realm' and 'public stability' to achieve 'solidity' and 'retrieve their sameness...their identity'. In Arendt this leads to the notion of 'solidarity' in a plural world. 'Hannah Arendt and feminist politics', in Shanley and Pateman, 235.

40 L. Code, *Rhetorical Spaces*, 132.

41 Co-authors L. Code, M. Ford, K. Martindale, S. Shewin and D. Shogan, *Is Feminist Ethics Possible?* (Ottawa: CRIAW/ICREF, 1991) 23.

42 S. Benhabib, 'The generalised and concrete other', in Fraser *et al.* 1992: 281.

43 H. Rose, *Love, Power and Knowlege*, 238.

44 P. Hill Collins, *Black Feminist Thought*, 207.

45 The number is alarmingly reduced to two, metaphor and metonymy, by Jakobson and the structuralists who followed him.

46 See P. Hill Collins, *Black Feminist Thought*, for the idea of beauty as a 'powerful alternative to Eurocentric aesthetics' could be a 'dual emphasis on beauty occurring via individual uniqueness juxtaposed in a community setting and on the importance of creating functional beauty from the scraps of everyday life' (89).

47 M. Frye (1987) 'The possibility of feminist theory', in A. Jaggar and P. Rothenburg (eds) *Feminist Frameworks: Alternative Theoretical Accounts of the Relations Between Women and Men* (London: McGraw Hill, 1993).

48 By consciousness-raising I do not refer, nor do I take Frye to refer, to the kind of activity described by N. Yuval-Davies (1997: 10) that works as a universalising process to persuade all women that they have common aims, but far more the 're-articulation' discussed by P. Hill Collins in which there is an affirmation of a 'consciousness that already exists' (1990: 30–1).

49 One of these communities consisted of the women at the University of Calgary who are concerned variously with Women's Studies and Canadian Literature,

who have provided generous intellectual and emotional support; and the other community is closer to home, and is represented largely by Margaret Beetham, at Manchester Metropolitan University, and the many people to whom she has introduced me.

50 J. Perreault, *Writing Selves*, 101.

51 L. Hunter and R. O'Rourke, *Creative Writing Strategies for English Studies* (Leeds: Adult and Continuing Education, 1996).

52 D. Haraway, *Modest Witness*. This book is directly pertinent to many issues raised in these discussions, but unfortunately was published after the submission of this text to my publisher. I comment here only to alert any reader to its importance, particularly to autography in science.

53 Hilary Rose performed this positioning in the paper given to the Women and Texts/Les Femmes et les Textes conference held in Leeds, July 1997. Quotations are from typescript.

54 N. Yuval-Davis, *Gender and Nation*, 88.

55 N. Brossard, 'Tender skin my mind', trans. Dympna Borowska, in Women and Words Collective, 180.

Bibliography

Aaron, J. and Walby, S. (eds) (1991) *Out of the Margins: Women's Studies in the Nineties*, London: The Falmer Press.

Adorno, T. (1982) *Against Epistemology: A Metacritique*, Oxford: Blackwell.

Althusser, L. (1971) 'Ideology and ideological state apparatuses (notes towards an investigation)', in *Lenin and Philosophy and Other Essays*, trans. B. Brewster, London: New Left Books.

Anzaldua, G. and Moraga, C. (eds) (1981) *This Bridge Called My Back: Writings by Radical Women of Colour*, Watertown MA: Persephone Press.

Angenot, M. (1988) 'Social discourse analysis: outlines of a research project', *Discourse Social / Social Discourse*, 1:1.

Apetheker, B. (1989) *Tapestries of Life: Women's work, women's Consciousness, and the Meaning of Daily Experience*, Amherst MA: University of Massachusetts Press.

Arendt, H. (1975/1951) *The Origins of Totalitarianism*, New York: Harcourt Brace Jovanovitch.

Aristotle (1984) *The Complete Works of Aristotle*, revised Oxford translation, ed. J. Barnes, Bollingen Series LXXI vols 1 and 2, Ithaca NY: Princeton University Press.

Attar, D. (1987) *Household Books Published in Britain, 1800–1914*, ed. L. Hunter, London: Prospect Books.

Austin, J. (1962) *How to Do Things with Words*, Oxford: Clarendon Press.

Bacon, F. (1974) *The Advancement of Learning*, in *Works of Bacon, VI*, Oxford: Clarendon Press.

Bauman, Z. (1989) *Modernity and Ambivalence*, Oxford: Polity Press.

——(1992) *Intimations of Postmodernity*, London: Routledge.

Bazerman, C. (1988) *Shaping Written Knowledge: The Genre and Activity of the Experimental Article in Science*, Madison WI: University of Wisconsin Press.

Beeman, W. *et al.* (1989) 'Hypertext and pluralism: from lineal to non-lineal thinking', *Proceedings of Hypertext '87*.

Beetham, D. (1991) *The Legitimation of Power*, London: Macmillan.

Beetham, D. and Boyle, K. (1995) *Introducing Democracy: 80 Questions and Answers*, Cambridge: Polity Press/UNESCO.

Beetham, M. (1996) *A Magazine of her Own: Domesticity and Desire in the Woman's Magazine 1800–1914*, London: Routledge.

——(1990) 'Towards a theory of the periodical as a publishing genre', in Brake, L., Jones, L. and Madden, L. (eds) *Investigating Victorian Journalism*, London: Macmillan.

Belenky, M. F., McVicker, C. B., Goldberger, N. R. and Tarule, J. M. (1986) *Women's Ways of Knowing: The Development of Self, Voice and Mind*, New York: Basic Books.

Benhabib, S. (1992) 'The generalized and the concrete other', in Fraser, E., Hornsby, J. and Lovibond, S., *Ethics: A Feminist Reader*, Oxford: Blackwell.

Benjamin, A., Cantor, G. and Christie, J. (eds) (1987) *The Figural and the Literal: Problems of Language in the History of Science and Philosophy, 1630–1800*, Manchester: Manchester University Press.

Bergendahl, G. (1990) 'Professional skill and traditions of knowledge', in Goranzon, B. and Florin, M. (eds) *Artificial Intelligence, Culture and Language: On Education and Work*, London: Springer Verlag.

Berlin, B. (1977) 'Speculations on the growth of ethnobotanic nomenclature', in Blount, B. and Sanchez, M. (eds) *Socio-cultural Dimensions of Language Change*, London: Academic Press.

Bernstein, R. (ed.) (1985) *Habermas and Modernity*, Cambridge: Cambridge University Press.

Birnbaum, L. (1991) 'Rigor mortis: a response to Nilsson's "Logic and artificial intelligence" ', *Artificial Intelligence*, 47.

Black, W. (1986/7) *Intelligent Knowledge-based Systems*, Wokingham: van Nostrand Rheinhold.

Bock, G. and James, S. (eds) (1992) *Beyond Equality and Difference: Citizenship, Feminist Politics and Female Subjectivity*, London: Routledge.

Boden, M. (1989a) 'Artificial intelligence: opportunities and dangers', in Murray, L. and Richardson, J. (eds) *Intelligent Systems in a Human Context: Development, Implications, and Applications*, London: Oxford University Press.

——(1989b) 'The meeting of man and machine', in Murray, L. and Richardson, J. (eds) *Intelligent Systems in a Human Context: Development, Implications, and Applications*, London: Oxford University Press.

——(1990) 'Artificial intelligence and images of man', in Ennals, R. and Gardin, J.-C. (eds) *Interpretation in the Humanities: Perspectives from Artificial Intelligence*, British Library, Library and Information Research Report 71, London: British Library.

Bolter, J. (1984) *Turing's Men: Western Culture in the Computer Age*, Chapel Hill NC: University of North Carolina Press.

Booth, W. (1974) *A Rhetoric of Irony*, Chicago IL: University of Chicago Press.

Boyce, G., Curran, J. and Wingate, P. (eds) (1978) *Newspaper History from the Seventeenth Century to the Present Day*, London: Constable.

Braidotti, R. (1992) 'On the female feminist subject, or: from "she-self" to "she-other" ', in Bock, G. and James, S. (eds) *Beyond Equality and Difference: Citizenship, Feminist Politics and Female Subjectivity*, London: Routledge.

Brooks, R. (1991) 'Intelligence with representation', *Artificial Intelligence*, 47, 1991.

Brown, V. (1994) *Adam Smith's Discourse: Canonicity, Commerce and Conscience*, London: Routledge.

Burnard, L. (1985) *Famulus Users' Manual*, Oxford University Computing Service.

Burrows, J. (1986a) *Computation into Criticism*, Oxford: Clarendon Press.

——(1986b) 'Modal verbs and moral principles', *JALLC*, 1:1.

——(1987) 'Word patterns and story shapes: the statistical analysis of narrative style', *JALLC*, 2:1.

Butler, J. (1993) *Bodies that Matter: On the Discursive Limits of 'Sex'*, London: Routledge.

Canovan, M. (1992) *Hannah Arendt: A Reinterpretation of her Political Thought*, Cambridge: Cambridge University press.

Cantor, G. (1989) 'The Rhetoric of experiment', in Gooding, D., Pinch, T. and Schaffer, S. (eds) *The Uses of Experiment: Studies in the Natural Sciences*, Cambridge: Cambridge University Press.

Cave, T. (1988) *Recognitions: A Study in Poetics*, Oxford: Clarendon Press.

Charney, D. (1989) 'Comprehending non-linear text: the role of discourse cues and reading strategies', *Proceedings of Hypertext '87*.

Charniak, E. and McDermott, D. (1985) *Introduction to Artificial Intelligence*, Wokingham: Addison Wesley.

Charolles, M. (1990) 'Logicist analysis and discourse analysis', in Ennals, R. and Gardin, J.-C. (eds) *Interpretation in the Humanities: Perspectives from Artificial Intelligence*, British Library, Library and Information Research Report 71, London: British Library.

Christensen, K. (1986) 'Technological decisions are moral decisions', in Geiss, G. and Viswanathan, N., *The Human Edge: Information Technology and Helping People*, New York: Haworth Press.

Churchland, P. M. and Churchland, P. S. (1990) 'Could a machine think?', *Scientific American*, 262:I.

Cicourel, A. (1964) *Method and Management in Sociology*, London: Macmillan.

Clancey, W. (1992) 'Model construction operators', *Artificial Intelligence*, 59.

——(1993) 'Notes on "heuristic classification" ', *Artificial Intelligence*, 59.

Code, L. (1995) *Rhetorical Spaces in Gendered Locations*, London: Routledge.

Code, L., Ford, M., Martindale, K., Shewin, S. and Shogan, D. (1991) *Is Feminist Ethics Possible?*, Ottawa: CRIAW/ICREF.

Collins, P. H. (1990) *Black Feminist Thought: Knowledge, Consciousness and Politics of Empowerment*, London: Harper Collins Academic.

Connerton, P. (1989) *How Societies Remember*, Cambridge: Cambridge University Press.

Cooley, M. (n.d.) *Architect or Bee? The Human/Technology Relationship*, comp. and ed. S. Cooley, Slough: Hand and Brain Publications.

——(1987) 'Creativity, skill and human-centred systems', in Goranzon, B. and Josefson, L. (eds) *Knowledge, Skill and Artificial Intelligence*, London: Springer Verlag.

——(1988) *Cognitive Science and its Applications for Human Computer Interaction*, London: Lawrence Erlbaum Associates.

Cook-Gumperz, J. (ed.) (1986) *The Social Construction of Literacy*, London: Cambridge University Press.

Cowan, J. and Sharp, D. (1988) 'Neural nets and artificial intelligence', in Graubard, S. (ed.) (1988) *The Artificial Intelligence Debate*, London: MIT Press.

Crump, M. (1989) 'Searching ESTC on BLAISE-LINE', *Factotum*, occasional paper 6.

Date, C. (1986) *An Introduction to Database Systems*, Wokingham: Addison-Wesley.

Dear, P. (ed.) (1991) *The Literary Structure of Scientific Argument*, Philadephia PA: University of Pennsylvania Press.

Deen, S. (1985) *Principles and Practice of Database Systems* London: Macmillan.

DeLauretis, T. (1984) *Alice Doesn't: Feminism, Semiotics, Cinema*, Bloomington IN: Indiana University Press.

——(1990) 'Eccentric subjects: feminist theory and historical consciousness', *Feminist Studies*, 19:1, spring.

Dennett, D. (1981) 'Where am I?', in Hofstadter, D. and Dennett, D. (eds) *The Mind's I: Fantasies and Reflections on Self and Soul*, Brighton: Harvester.

——(1988) 'When philosophers encounter artificial intelligence', in Graubard, S. (ed.) *The Artificial Intelligence Debate*, London: MIT Press.

Derrida, J. (1974) *Of Grammatology*, trans. G. Spivak, London: Johns Hopkins University Press.

Diamond, I. and Quinby, L. (eds) (1988) *Feminism and Foucault: Reflections on Resistance*, Boston MA: Northeastern University Press.

Dietz, M. (1991) 'Hannah Arendt and feminist politics', in Shanley, M. L. and Pateman, C. (eds) *Feminist Interpretations and Political Theory*, Cambridge: Polity Press.

DiLeonardo, M. (1991) 'Contingencies of value in feminist anthropology', in Hartman, J. and Messer-Davidow, E. (eds) *(En) Gendering Knowledge: Feminists in Academe*, Knoxville TN: University of Tennessee.

DiStephano, C. (1991) 'Masculine Marx', in Shanley, M. L. and Pateman, C. (eds) *Feminist Interpretations and Political Theory*, Cambridge: Polity Press.

Doell, R. (1991) 'Whose research is this? Values and biology', in Hartman, J. and Messer-Davidow, E. (eds) *(En) Gendering Knowledge: Feminists in Academe*, Knoxville TN: University of Tennessee.

Doran, J. (1989) 'Distributed artificial intelligence and the modelling of socio-cultural systsems', in Murray, L. and Richardson, J. (eds) *Intelligent Systems in a Human Context: Development, Implications and Applications*, London: Oxford University Press.

Dreyfus, H. (1981) 'From micro-worlds to knowledge representation: AI at an impasse', in Haugeland, J. (ed.) *Mind Design: Philosophy, Psychology, Artificial Intelligence* London: MIT Press.

——(1988) 'The Socratic and Platonic basis of cognitivism', *AI and Society*, 2:107.

Dreyfus, H. and Dreyfus, S. (1988) 'Making a mind versus modeling the brain: artificial intelligence back at a branchpoint', in Graubard, S. (ed.) *The Artificial Intelligence Debate*, London: MIT Press.

Driver, E. (1989) *Cookery Books Published in Britain, 1875–1914*, ed. L. Hunter, London: Prospect Books.

Duncan, K. and Harris, D. (eds) (1985) *Computers in Education*, Proc. IFIP TC3, 4th World Conference on Computers in Education, North Holland: Elsevier.

Eagleton, T. (1990) 'Nationalism, irony and commitment', in *Nationalism, Colonialism and Literature*, intro. S. Deane, Minneapolis MN: University of Minnesota Press.

Eamon, W. (1994) *Science and the Secrets of Nature: Books of Secrets in Medieval and Early Modern Culture*, Ithaca NY: Princeton University Press.

Edwards, D. (1989) ' "Lost in hyperspace": cognitive mapping and navigation in a hypertext environment', in McAleese, R. (ed.) *Hypertext: Theory into Practice*, Oxford: Intellect Books.

Ellul, J. (1980) 'The power of technique and the ethics of non-power', in Woodward, K. (ed.) *The Myths of Information: Technology and Post-industrial Culture*, London: Routledge and Kegan Paul.

Elshtain, J. (1992) 'The power and powerlessness of women', in Bock, G. and James, S. (eds) *Beyond Equality and Difference: Citizenship, Feminist Politics and Female Subjectivity*, London: Routledge.

Elsom-Cook, M. (1991a) 'Multimedia: the emperor's new clothes', *ALTNEWS, New Technologies and Learning in Europe*, 6, February.

——(1991b) *Artificial Intelligence and Human Institutions*, London: Springer Verlag.

Ennals, R. (1987) 'Humanities and computing', in Rahtz, S. (ed.) *Information Technology in the Humanities: Tools, Techniques and Applications*, Chichester: Ellis Horwood.

——(1990) 'Interpretation and codebreaking', in Ennals, R. and Gardin, J-C. (eds) (1990) *Interpretation in the Humanities: Perspectives from Artificial Intelligence*, British Library, Library and Information Research Report 71, London: British Library.

Ennals, R. (ed.) (1987) *Artificial Intelligence, State of the Art Report*, 15:3. Oxford: Pergamon Infotech.

Ennals, R. and Gardin, J-C. (eds) (1990) *Interpretation in the Humanities: Perspectives from Artificial Intelligence*, British Library, Library and Information Research Report 71, London: British Library.

Erickson, B. and Nosanchuk, T. (1983) *Understanding Data*, Milton Keynes: Open University Press.

Faigley, L. (1992) *Fragments of Rationality: Postmodernity and the Subject of Composition*, London: University of Pittsburgh Press.

Fargue, J. (1990) 'Remarks on the interrelations between artificial intelligence, mathematical logic and humanities', in Ennals, R. and Gardin, J.-C. (eds) *Interpretation in the Humanities: Perspectives from Artificial Intelligence*, British Library, Library and Information Research Report 71, London: British Library.

Felski, R. (1989) *Beyond Feminist Aesthetics: Feminist Literature and Social Change*, Cambridge MA: Harvard University Press.

Ferguson, K. (1993) *The Man Question: Visions of Subjectivity in Feminist Theory*, Oxford: University of California Press.

Feyerabend, P. (1988) [1975] *Against Method*, London: Verso.

Finnegan, R., Salaman, G. and Thompson, K. (eds) (1987) *Information Technology: Social Issues, a Reader*, London: Hodder and Stoughton and the Open University.

Flax, J. (1983) 'Political philosophy and the patriarchal unconscious', in Harding, S. and Hintikka, M. (eds) (1983) *Discovering Reality: Feminist Perspectives on Epistemology, Methodology and Philosophy of Science*, Dordrecht: Reidel.

——(1990) *Thinking in Fragments: Psychoanalysis, Feminism, and Postmodernism in the Contemporary West*, Berkeley CA: University of California Press.

——(1992) 'Beyond equality: gender, justice and difference', in Bock, G. and James, S. (eds) *Beyond Equality and Difference: Citizenship, Feminist Politics and Female Subjectivity*, London: Routledge.

——(1993) *Disputed Subjects: Essays on Psychoanalysis, Politics and Philosophy*, London: Routledge.

Fodor, J. (1981) 'Methodological solipsism considered as a research strategy in cognitive psychology', in Haugeland, J. (ed.) *Mind Design: Philosophy, Psychology, Artificial Intelligence*, London: MIT Press.

Foucault, M. (1970) *The Order of Things*, London: Tavistock.

——(1979) *Language, Counter-memory, Practice*, ed. D. Bouchard, Ithaca NY: Cornell University Press.

Fox-Genovese, E. (1996) *Feminism is not the Story of My Life: How Today's Feminist Elite has Lost Touch with the Real Concerns of Women*, New York: Nan A. Talese.

Fox-Keller, E. (1982) 'Feminism and science', *Signs*, 7:3.

——(1992) *Secrets of Life, Secrets of Death: Essays on Language, Gender and Science*, London: Routledge.

Fox-Keller, E. and Longino, H. (eds) (1996) *Feminism and Science*, Oxford: Oxford University Press.

Fraser, E., Hornsby, J. and Lovibond, S. (1992) *Ethics: A Feminist Reader*, Oxford: Blackwell.

Fraser, N. (1989) *Unruly Practices: Power, Discourse and Gender in Contemporary Social Theory*, Cambridge: Polity Press.

——(1991) 'What's critical about critical theory? The case of Habermas and gender', in Shanley, M. L. and Pateman, C. (eds) *Feminist Interpretations and Political Theory*, Cambridge: Polity Press.

Fricker, M. (1994) 'Knowledge as construct: theorizing the role of gender in knowledge', in Lennon, K. and Whitford, M. *Knowing the Difference: Feminist Perspectives in Epistemology*, London: Routledge.

Frude, N. (1989) 'Intelligent systems off the shelf: the high street consumer and artificial intelligence', in Murray, L. and Richardson, J. (eds) *Intelligent Systems in a Human Context: Development, Implications and Applications*, London: Oxford University Press.

Frye, M. (1993) [1987] 'The possiblity of feminist theory', in Jaggar, A. and Rothenburg, P. *Feminist Frameworks: Alternative Theoretical Accounts of the Relations Between Women and Men*, London: McGraw Hill.

Gadamer, H-G. (1981) [1976] *Reason in the Age of Science*, trans. F. Lawrence, London: MIT Press.

Gardin, J.-C. (1990) 'Interpretation in the humanities: Some thoughts on the third way' in Ennals, R. and Gardin, J.-C. (eds*)* *Interpretation in the Humanities: Perspectives from Artificial Intelligence*, British Library, Library and Information Research Report 71, London: British Library.

Garver, E. (1988) 'Aristotle's *Rhetoric* on unintentionally hitting the principles of the sciences', *Rhetorica*, VI:4, autumn.

Geiss, G. and Viswanathan, N. (1986) *The Human Edge: Information Technology and Helping People*, New York: Haworth Press.

Gellner, E. (1982) 'Nationalism and the two forms of cohesion in complex societies', *Proceedings of the British Academy*, LXVIII, London: Oxford University Press.

Giddens, A. (1980) [1973] *The Class Structure of the Advanced Societies*, London: Hutchinson.

——(1981) *A Contemporary Critique of Historical Materialism Vol. 1: Power, Property and the State*, London: Macmillan.

——(1987) 'Structuralism, post-structuralism and the production of culture', in Giddens, A. and Turner, J. (eds) *Social Theory Today*, Cambridge: Polity Press.

Gill, K. (1987) 'Artificial intelligence and social action: education and training', in Goranzon, B. and Josefson, L. (eds) *Knowledge, Skill and Artificial Intelligence*, London: Springer Verlag.

Gilligan, C. (1982) *In a Different Voice: Psychological Theory and Women's Development*, Cambridge MA: Harvard University Press.

Golinski, J. (1982) *Science as public culture*, Cambridge: Cambridge University Press.

——(1990) 'The theory of practice andthe practice of theory: sociological approaches in the history of science', *Isis*, 81.

Gooding, D. (1989) ' "Magnetic curves" and the magnetic field: experimentation and representation in the history of a theory', in Gooding, D., Pinch, T. and Schaffer, S. (eds) *The Uses of Experiment: Studies in the Natural Sciences*, Cambridge: Cambridge University Press.

Gooding, D., Pinch, T. and Schaffer, S. (eds) (1989) *The Uses of Experiment: Studies in the Natural Sciences*, Cambridge: Cambridge University Press.

Goranzon, B. (1987) 'The practice of the use of computers: a paradoxical encounter between different traditions of knowledge', in Goranzon, B. and Josefson, L. (eds) *Knowledge, Skill and Artificial Intelligence*, London: Springer Verlag.

Goranzon, B. and Florin, M. (eds) (1990) *Artificial Intelligence, Culture and Language: On Education and Work*, London: Springer Verlag.

Goranzon, B. and Josefson, L. (eds) (1987) *Knowledge, Skill and Artificial Intelligence*, London: Springer Verlag.

Goyet, F. (1991) 'The word "Commonplaces" in Montaigne', in Hunter, L. (ed.) *Topos, Commonplace and Cliché: Toward an Understanding of Analogical Reasoning*, London: Macmillan.

Grafton, A. and Jardine, L. (1986) *From Humanism to the Humanities*, London: Duckworth.

Graubard, S. (ed.) (1988) *The Artificial Intelligence Debate*, London: MIT Press.

Greene, J. (1987) *Memory, Thinking and Language*, London: Methuen.

Griffiths, M. (1988) 'Strong feelings about computers', *Women's Studies International Forum*, 11:2.

Habermas, J. (1987) *The Philosophical Discourse of Modernity*, Cambridge: Polity Press.

——(1985) 'Reply to Rorty', in Bernstein, R. (ed.) *Habermas and Modernity*, Cambridge: Cambridge University Press.

——(1973) [1971] *Theory and Practice*, trans. J. Viertel, Boston MA: Beacon Press.

Halliday, M. and Hasan, R. (1976) *Cohesion in English*, New York: Longman.

Hanen, M. and Nielsen, K. (1987) 'Science, morality and feminist theory', *Canadian Journal of Philosophy*, 13.

Haraway, D. (1988) 'Situated knowledges: the science quesion in feminism and the privilege of partial perspective', *Feminist Studies*, 14:3.

——(1990) [1985] 'A manifesto for cyborgs: science, technology and socialist feminism in the 1980s', in Nicholson, L. (ed.) *Feminism/Postmodernism*, London: Routledge.

——(1997)*Modest_Witness@Second_Millenium.FemaleMan©_Meets_Onco-Mouse*^{TM}, London: Routledge.

Harding, S. (1986) *The Science Question in Feminism*, Milton Keynes: Open University Press.

——(1991) *Whose Science? Whose Knowledge? Thinking from Women's Lives*, Milton Keynes: Open University Press.

Harding, S. and Hintikka, M. (eds) (1983) *Discovering Reality: Feminist Perspectives on Epistemology, Methodology and Philosophy of Science*, Dordrecht: Reidel.

Hardman, L. and Edwards, D. (1989) ' "Lost in hyperspace": cognitive mapping and navigation in a hypertext environment', in McAleese, R. (ed.) *Hypertext: Theory into Practice*, Oxford: Intellect Books.

Harnad, S. (1989) 'Minds, machines and Searle', *Journal of Experimental and Theoretical Artificial Intelligence*, 1:I.

Hartman, J. and Messer-Davidow, E. (eds) (1991) *(En) Gendering Knowledge: Feminists in Academe*, Knoxville TN: University of Tennessee.

Harwood, J. (1991) *Robert Boyle: The Early Essays and Ethics of Robert Boyle*, Carbondale IL: Southern Illinois University Press.

Haugeland, J. (1985) *Artificial Intelligence: The Very Idea*, London: MIT Press.

Haugeland, J. (ed.) (1981) *Mind Design: Philosophy, Psychology, Artificial Intelligence*, London: MIT Press.

Havelock, E. (1963) *Preface to Plato*, Cambridge MA: Harvard University Press.

Hekman, S. (1990) *Gender and Knowledge: Elements of a Postmodern Feminism*, Oxford: Polity Press/Blackwell.

Henriquez, J., Hollway, W., Urwin, C., Venn, C. and Walkerdine, V. (1984) *Changing the Subject: Psychology, Social Regulation and Subjectivity*, London: Methuen.

Hill Collins, P. (1990) *Black Feminist Thought: Knowledge, Consciousness, and the Politics of Empowerment*, London: Routledge.

Hobsbawm, E. and Ranger, T. (eds) (1983) *The Invention of Tradition*, Cambridge: Cambridge University Press.

Hockey, S. (1980) *Guide to Computer Applications in the Humanities*, London: Duckworth.

Hockey, S. and Martin, J. (1988) *Oxford Concordance Program*, Oxford: Oxford University Computing Service.

Hofstadter, D. (1979) *Gödel, Escher, Bach: An Eternal Golden Braid*, Harmondsworth: Penguin.

Hofstadter, D. and Dennett, D. (eds) (1981) *The Mind's I: Fantasies and Reflections on Self and Soul*, Brighton: Harvester.

Howell, W. (1956) *Logic and Rhetoric in England, 1500–1700*, Princeton NJ: Princeton University Press.

Huizinga, J. (1949) [1944] *Homo Ludens: A Study of the Play-elements in Culture*, London: Routledge and Kegan Paul.

Hunter, L. (1984a) *Rhetorical Stance in Modern Literature*, London: Macmillan.

——(1984b) *George Orwell: The Search for a Voice*, Milton Keynes: Open University Press.

——(1989) *Modern Allegory and Fantasy*, London: Macmillan.

——(1990a) 'The computer as machine: friend or foe?' in Turk, C. R. R. (ed.) *Humanities Research Using Computing*, London: Chapman and Hall.

——(1990b) 'Creating a hypertext with guide', in Darby, J. (ed.) *The CTISS File*, 9, February.

——(1990c) 'Fact-information-data-knowledge: databases as a way of organizing knowledge', *Journal of the Association for Literary and Linguistic Computing*, 5:1.

——(1990d) 'A rhetoric of mass communication: collective or corporate public discourse', in Enos, R. (ed.) *Oral and Written Communication: Historical approaches*, London: Sage.

——(1991a) 'Remember Frankenstein: rhetoric and artificial intelligence', *Rhetorica*, IX:4, winter.

——(1991b) 'Watson and McLuhan's *From Cliche to Archetype*', in Hunter, L. (ed.) *Topos, Commonplace and Cliché: Toward an Understanding of Analogical Reasoning*, London: Macmillan.

——(1993) 'AI and representation: a study of a rhetorical context for legitimacy', *AI and Society*, 7.

——(1994) 'The socialising of context: methodologies for hypertext', in Hunter, L., Beetham, M., Fuller, D. *et al. The Victorian Periodicals Project*, Oxford: CTI Centre for Texts.

——(1996a) 'Bodily functions in Cartesian space', in Chew, S. and Hunter, L. (eds) *Borderblur: Poetry and Poetics in Canadian Literature*, Edinburgh: Quadriga.

——(1996b) *Outsider Notes: Feminist Approaches to the Nation State, Writers/Readers and Publishing in Canada 1960–95*, Vancouver: Talonbooks.

——(1997) 'Household books, 1500–1700', in Sedge, D. (ed.) *Cambridge Bibliography to English Literature, 1500–1700, third edition*, Cambridge: Cambridge University Press.

Hunter, L. (ed.) (1991) *Topos, Commonplace and Cliché: Toward an Understanding of Analogical Reasoning*, London: Macmillan.

Hunter, L. and Hutton, S. (eds) (1997) *Women in Medicine, Science and Technology, 1500–1700*, Stroud: Alan Sutton.

Hunter, L., Beetham, M., Fuller, D. *et al.* (1994) *The Victorian Periodicals Project*, Oxford: CTI Centre for Texts.

Hunter, L. and O'Rourke, R. (1996) *Creative Writing Strategies for English Studies*, Leeds: Adult and Continuing Education.

Jahnke, H. and Otte, M. (eds) (1981) *Epistemological and Social Problems of the Science in the Early Nineteenth Century*, London and Dordrecht: Riedel.

Jacob, P. (1990) 'What is interpretation?: a philosophical view' in Ennals, R. and Gardin, J.-C. (eds) *Interpretation in the Humanities: Perspectives from Artificial Intelligence*, British Library, Library and Information Research Report 71, London: British Library.

Jacobus, M., Fox-Keller, E. and Shuttleworth, S. (1990) *Body Politics: Women and the Discourses of Science*, London: Routledge.

Jaggar, A. (1983) *Feminist Politics and Human Nature*, Brighton: Harvester.

——(1989) 'Love and knowledge: emotion in feminist epistemology', in Jaggar, A. and Bordo, S., *Gender/Body/Knowledge: Feminist Deconstructions of Being and Knowing*, New Brunswick NJ: Rutgers University Press.

Jaggar, A. and Bordo, S. (1989) *Gender/Body/Knowledge: Feminist Deconstructions of Being and Knowing*, New Brunswick NJ: Rutgers University Press.

Jaggar, A. and Rothenburg, P. (1993) *Feminist Frameworks: Alternative Theoretical Accounts of the Relations Between Women and Men*, London: McGraw Hill.

Jakobson, R. (1956) 'Two aspects of language and two types of aphasic disturbances', in Jakobson, R. and Halle, M. *Fundamentals of Language*, The Hague: Mouton.

James, S. (1992) 'The good-enough citizen: female citizenship and independence', in Bock, G. and James, S. (eds) *Beyond Equality and Difference: Citizenship, Feminist Politics and Female Subjectivity*, London: Routledge.

Jameson, F. (1986) [1979] 'Foreword', in Lyotard, J., *The Postmodern Condition: A Report on Knowledge*, trans. G. Bennington and B. Massumi, Manchester: Manchester University Press.

Janik, A. (1987) 'Tacit knowledge, working life and scientific "method" ', in Goranzon, B. and Josefson, L. (eds) *Knowledge, Skill and Artificial Intelligence*, London: Springer Verlag.

——(1990) 'Tacit knowledge, rule-following and learning', in Goranzon, B. and Florin, M. (eds) *Artificial Intelligence, Culture and Language: On Education and Work*, London: Springer Verlag.

Jardine, L. (1994) *Erasmus, Man of Letters*, London: Cambridge University Press.

Jarratt, S. (1990) 'The first sophists and feminism: discourses of the "Other" ', *Hypatia*, 5:1.

——(1991) *ReReading the Sophists: Classical Rhetoric Refigured*, Carbondale and Edwardsville IL: Southern Illinois Press.

Johns, A. (1991) 'History, science, and the history of the book: The making of natural philosophy in early modern England', *Publishing History*, 30.

Jonassen, D. (1989) *Hypertext/Hypermedia*, Englewood Cliffs NJ: Prentice Hall.

Johannessen, K. (1990) 'Rule-following and intransitive understanding', in Goranzon, B. and Florin, M. (eds) *Artificial Intelligence, Culture and Language: On Education and Work*, London: Springer Verlag.

Johnson-Laird, P. (1989) 'Human experts and expert system', in Murray, L. and Richardson, J. (eds) *Intelligent Systems in a Human Context: Development, Implications and Applications*, London: Oxford University Press.

Kellas, J. (1991) *The Politics of Nationalism and Ethnicity*, London: Macmillan.

Kirsch, D. (1991a) 'Foundations of AI: the big issues', *Artificial Intelligence*, 47.

——(1991b) 'Today the earwig, tomorrow man?', *Artificial Intelligence*, 47.

Kuhn, T. (1962) *The Structure of Scientific Revolutions*, Chicago IL: Chicago University Press.

Landow, G. (1992) *Hypertext: The Convergence of Contemporary Critical Theory and Technology*, London: Johns Hopkins University Press.

Latour, B. (1989) 'Clothing the naked truth', in Lawson, H. and Appignanesi, L. (eds) *Dismantling Truth: Reality in the Post-modern World*, London: Weidenfeld and Nicolson.

Latour, B. and Woolgar, S. (1979) *Laboratory Life: The Social Construction of Scientific Facts*, Beverley Hills CA: Sage.

Lawson, H. (1985) 'Addressing fundamental problems in computer-related education and training', in Duncan, K. and Harris, D. (eds) *Computers in Education*, Proc. IFIP TC3, 4th World Conference on Computers in Education, North Holland: Elsevier.

Lawson, H. and Appignanesi, L. (eds) (1989) *Dismantling Truth: Reality in the Post-modern World*, London: Weidenfeld and Nicolson.

Lechner, J. (1962) *Renaissance Concepts of the Commonplaces*, Westport CT: Greenwood.

Lenat, D. and Guha, R. (1989) *Building Large Knowlege-based Systems*, Reading MA: Addison Wesley.

Lennon, K. and Whitford, M. (eds) (1994) *Knowing the Difference: Feminist Perspectives in Epistemology*, London: Routledge.

Leslie, M. (1990) 'The Hartlib Papers Project: text retrieval with large datasets', *JALLC*, 5:1.

Lloyd, G. (1984) *The Man of Reason: 'Male' and 'Female' in Western Philosophy*, London: Methuen.

Longino, H. (1996) 'Subjects, Power, and Knowledge: Description and Prescription in Feminist Philosophies of Science', in Fox-Keller, E. and Longino, H. (eds) *Feminism and Science*, Oxford: Oxford University Press.

Lovegrove, G. and Segal, B. (eds) (1991) *Women into Computing: Selected Papers 1988–90*, London: Springer Verlag.

Lovibond, S. (1994) 'The end of morality?' in Lennon, K. and Whitford, M. (eds) *Knowing the Difference: Feminist Perspectives in Epistemology*, London: Routledge.

Lyotard, J. (1986) [1979] *The Postmodern Condition: A Report on Knowledge*, trans. G. Bennington and B. Massumi, fwd by F. Jameson, Manchester: Manchester University Press.

McAleese, R. (ed.) (1989) *Hypertext: Theory into Practice*, Oxford: Intellect Books.

McCarthy, J. (1988) 'Mathematical logic in artificial intelligence', in Graubard, S. (ed.) *The Artificial Intelligence Debate*, London: MIT Press.

McDermott, D. (1981) 'Artificial intelligence meets natural stupidity', in Haugeland, J. (ed.) *Mind Design: Philosophy, Psychology, Artificial Intelligence*, London: MIT Press.

Macdonell, D. (1986) *Theories of Discourse*, Oxford: Blackwell.

McGuire, J. and Melia, T. (1989) 'Some cautionary strictures on the writing of the rhetoric of science', *Rhetorica*, VII:1.

Maddison, R. (1969) *The Life of the Hon. Robert Boyle*, London: Taylor and Francis.

Machievelli, N. (1992) *The Prince*, New York: Dover.

Mandel, H. and Levin, J. (eds) (1988) *Knowledge Acquisition from Texts and Pictures*, Amsterdam: Elsevier.

Medawar, P. (1986) [1964] 'Is the scientific paper a fraud?' in Brown, J., Cooper, A., Huton, T., Toates, F. and Zeldin, D. (eds) *Science in Schools*, Milton Keynes: Open University Press.

Medina, V. (1990) *Social Contract Theories: Political Obligation or Anarchy?*, Savage MD: Rowman and Littlefield Publishers.

Merchant, C. (1980) *The Death of Nature: Women, Ecology and the Scientific Revolution*, London: Wildwood House.

Miall, D. (1990) 'An expert system approach to the interpretation of literary structure', in Ennals, R. and Gardin, J.-C. (eds) *Interpretation in the Humanities: Perspectives from Artificial Intelligence*, British Library, Library and Information Research Report 71, London: British Library.

Michie, D. and Johnston, R. (1985) [1984] *The Creative Computer: Machine Intelligence and Human Knowledge*, Harmondsworth: Penguin.

Miles, I., Rush, H., Turner, K. and Bessant, J. (1988) *Information Horizons: The Long-term Social Implications of New Information Technologies*, Aldershot: Edward Elgar.

Minsky, M. (1981) 'A framework for presenting knowledge', in Haugeland, J. (ed.) *Mind Design: Philosophy, Psychology, Artificial Intelligence*, London: MIT Press.

Mohanty, C., Russo, A. and Torres, L. (1991) *Third World Women and the Politics of Feminism*, Indianapolis IN: Indiana University Press.

Molina, A. (1989) *The Social Basis of the Microelectronics Revolution*, Edinburgh: Edinburgh University Press.

Moreton, A. (1978) *Literary Detection*, London: Scribner.

Moss, J. Dietz (1989) 'The interplay of science and rhetoric in seventeenth century Italy', *Rhetorica*, VII:1, winter.

Murphy, J. J. (ed.) (1982) *The Rhetorical Tradition and Modern Writing*, New York: Modern Languages Association.

Murray, L. and Richardson, J. (eds) (1989) *Intelligent Systems in a Human Context: Development, Implications and Applications*, London: Oxford University Press.

Mylopoulous, J. and Brodie, M. (eds) (1989) *Readings in Artificial Intelligence and Databases*, San Mateo CA: Morgan Kaufmann.

Nairn, T. (1977) *The Break-up of Britain: Crisis and Neo-Nationalism*, London: New Left Books.

Neel, J. (1994) *Aristotle's Voice*, Carbondale and Edwardsville IL: University of Illinois Press.

Nilsson, N. (1982) *Principles of Artificial Intelligence*, New York.

Nordenstam, T. (1987) 'Language and action', in Goranzon, B. and Florin, M. (eds) (1990) *Artificial Intelligence, Culture and Language: On Education and Work*, London: Springer Verlag.

Norman, N. (1990) 'The electronic teaching theatre: interactive hypermedia and mental models of the classroom', in Ward, P. (ed.) 'Hypermedia and artificial intelligence', special edition of *Current Psychology*, New Brunswick NJ: Rutgers University Press.

Olson, D. and Hildyard, A. (1983) 'Writing and literal meaning', in Martlew, M. (ed.) *The Psychology of Written Languages, Developmental and Educational Perspectives*, Chichester: Wiley.

Ong, W. (1971) *Rhetoric, Romance and Technology: Studies in the Interaction of Expression and Culture*, Ithaca NY: Cornell University Press.

O'Shea, T., Self, J. and Thomas, G. (eds) (1987) *Intelligent Knowledge-based Systems: An Introduction*, London: Harper & Row.

O'Shea, T. (1987) 'IKBS – setting the scene', in O'Shea, T., Self, J. and Thomas, G. (eds) *Intelligent Knowledge-based Systems: An Introduction*, London.

Papert, S. (1988) 'One AI or many?' in Graubard, S. (ed.) *The Artificial Intelligence Debate*, London: MIT Press.

Parke, M. (1976) 'Influence of concepts of *Ordinatio* and *compilatio* in the development of the book', in Alexander, J. and Gibson, M., *Medieval Learning*, Oxford: Clarendon Press.

Pateman, C. (1985) [1979] *The Problem of Political Obligation: A Critique of Liberal Theory*, Cambridge: Polity Press.

——(1991a) *Feminist Interpretations and Political Theory*, Cambridge: Polity Press.

——(1991b) ' "God hath ordained to man a helper": Hobbes, patriarchy and conjugal right', in Shanley, M. L. and Pateman, C. (eds) *Feminist Interpretations and Political Theory*, Cambridge: Polity Press.

——(1995) *Democracy, Freedom and Special Rights*, Swansea: University of Wales.

Pelling, M. (1994) 'Apprenticeship, health and social cohesion in early modern London', *History Workshop Journal*, 37.

Penrose, R. (1989) *The Emperor's New Mind: Concerning Computers, Minds and the Laws of Physics*, fwd by M. Gardner, London: Oxford University Press.

Perelman, C. and Olbrechts-Tyteca, L. (1971) *The New Rhetoric: A Treatise on Argumentation*, trans. J. Wilkinson and P. Weaver, London: Kluwer.

Perelman, M. (1990) *Information, Social Relations and the Economics of High Technology*, London: Macmillan.

Perreault, J. (1995) *Writing Selves: Contemporary Feminist Autography*, Minneapolis MN: University of Minnesota Press.

Perry, R. (1991) 'Introduction', *Signs*, 16:1.

Plato (1973a) *Phaedrus*, trans. W. Hamilton, Harmondsworth: Penguin.

Plato (1973b) *The Republic*, trans. D. Lee, Harmondsworth: Penguin.

Pocock, J. (1971) *Politics, Language and Time: Essays on Political Thought and History*, London: Methuen.

Polanyi, M. (1967) *The Tacit Dimension*, London: Routledge and Kegan Paul.

Poulakos, J. (1995) *Sophistical Rhetoric in Classical Greece*, Columbia SC: University of South Carolina Press.

Poulantzas, N. (1978) *State, Power, Socialism*, trans. P. Camiller, London: New Left Books.

Pravitz, D. (1987) 'Tacit knowledge – an impediment for AI?', in Goranzon, B. and Florin, M. (eds) (1990) *Artificial Intelligence, Culture and Language: On Education and Work*, London: Springer Verlag.

Rahtz, S. (1987a) 'The "processing" of words', in Rahtz, S. (ed.) *Information Technology in the Humanities: Tools, Techniques and Applications*, Ellis Horwood.

Rahtz, S. (ed.) (1987b) *Information Technology in the Humanities: Tools, Techniques and Applications*, Chichester: Ellis Horwood.

Ramzouoglu, C. (ed.) (1993) *Up Against Foucault: Explorations of Some Tensions between Foucault and Feminism*, London: Routledge.

Reynolds, N. (1993) 'Ethos as location: new sites for understanding discursive authority', *Rhetoric Review*, 11:2, spring.

Roberts, R. and Goode, J. (eds) (1993) *The Rediscovery of Rhetoric*, London: Bristol Classical Press.

Rorty, R. (1989) *Contingency, Irony, and Solidarity*, Cambridge: Cambridge University Press.

——(1991) *Essays on Heidegger and Others*, Cambridge: Cambridge University Press.

——(1985) 'Habermas and Lyotard on postmodernity', in Bernstein, R. (ed.) *Habermas and Modernity*, Cambridge: Cambridge University Press.

——(1991) *Objectivity, Relativism, and Truth*, Cambridge: Cambridge University Press.

Rose, H. (1983) 'Hand, brain and heart: towards a feminist epistemology for the natural sciences', *Signs*, 9:1.

——(1993) 'Rhetoric, feminism and scientific knowledge: or, from either/or to both/and', in Roberts, R. and Goode, J. (eds) *The Rediscovery of Rhetoric*, London: Bristol Classical Press.

——(1994) *Love, Power and Knowledge: Towards a Feminist Transformation of the Sciences*, London: Polity Press.

—— (1997a) 'Goodbye truth: hallo trust', in Maynared, M. (ed.) *Science and the Construction of Women*, London: University College London Press.

——(1997b) 'Subjectivity and sequences: moving beyond a determining culture', paper given to the Women and Texts/Les Femmes et les Textes, conference, Leeds, July 1997.

Rose, H. and Rose, S. (eds) (1976a) *The Political Economy of Science*, London: Macmillan.

——(1976b) *The Radicalisation of Science: Ideology of/in the Natural Sciences*, London: Macmillan.

——(1976c) 'The incorporation of science', in Rose, H. and Rose, S. (eds) *The Radicalisation of Science: Ideology of/in the Natural Sciences*, London: Macmillan.

Rose, S. (1973) 'Can science be neutral?', *Proceedings of the Royal Institute*, 45.

——(1992) *The Making of Memory: From Molecule to Mind*, London: Bantam Press.

Rosser, S. (1992) *Biology and Feminism: A Dynamic Interaction*, Oxford: Twayne.

Roszak, T. (1986) *The Cult of Information: The Folklore of Computing and the True Art of Thinking*, Cambridge: Lutterworth Press.

Rotblat, J. (ed.) (1982) *Scientists, the Arms Race and Disarmament*, London: Taylor and Francis.

Rouse, J. (1987) *Knowledge and Power. Toward a Political Philosophy of Science*, London: Cornell University Press.

Rowe, W. and Schelling, V. (1991) *Memory and Modernity: Popular Culture in Latin America*, London: Verso.

Sawicki, J. (1991) 'Foucault and feminism: toward a politics of difference', in Shanley, M. L. and Pateman, C. (eds) *Feminist Interpretations and Political Theory*, Cambridge: Polity Press.

Scanlon, R. (1966) [1958] 'Adolph Hitler and the technique of mass brainwashing', in *Rhetorical Idiom: Essays in Rhetoric, Oratory, Language and Drama*, New York: Russell and Russell.

Scheman, N. (1991) 'Who wants to know: the epistemological value of values', in Hartman, J. and Messer-Davidow, E. (eds) *(En)Gendering Knowledge: Feminists in Academe*, Knoxville TN: University of Tennessee Press.

Schiebinger, L. (1989) *The Mind has no Sex? Women in the Origins of Modern Science*, Cambridge MA: Harvard University Press.

Scribner, S. and Cole, M. (1981) *The Psychology of Literature*, London: Harvard University Press.

Searle, J. (1980) 'Minds, brains and programs', *Behavioural and Brain Sciences*, 3.

——(1987) 'Cognitive science and the computer metaphor', in Goranzon, B. and Florin, M. (eds) (1990) *Artificial Intelligence, Culture and Language: On Education and Work*, London: Springer Verlag.

——(1990) 'Is the brain's mind a computer program?', *Scientific American*, 262:1.

Shanley, M. L. and Pateman, C. (eds) (1991) *Feminist Interpretations and Political Theory*, Cambridge: Polity Press.

Shapin, S. (1988) 'The house of experiment', *Isis*, 79.

——(1991) ' "A scholar and a gentleman": the problematic identity of the scientific practitioner in early modern England', *Journal of the History of Science*, XXIX.

Sharples, M., Hogg, D., Hutchinson, C., Torrance, S. and Young, D. (eds) (1989) *Computers and Thought: A Practical Introduction to Artificial Intelligence*, London: MIT Press.

Slack, P. (1979) 'Mirrors of health and treasures of poor men', in Webster, C. (ed.) *Heath Medicine and Mortality in the Sixteenth Century*, Cambrdige: Cambridge University Press.

Slesnick, T. (1985) 'Software for girls: a sexist solution', in Duncan, K. and Harris, D. (eds) *Computers in Education*, Proc. IFIP TC3, 4th World Conference on Computers in Education, North Holland: Elsevier.

Sloan, T. (1974) 'The crossing of rhetoric and poetry in the English renaissance', in Sloan, T. and Waddington, C. (eds) *The Rhetoric of Renaissance Poetry from Wyatt to Milton*, London: University of California Press.

Sloman, A. (1992) 'The Emperor's new mind', *Artificial Intelligence*, 56.

Smith, B. (1991) 'The owl and the electric encyclopedia', *Artificial Intelligence*, 47.

236   *Bibliography*

Smith, D. (1987) *The Everyday World as Problematic*, Boston: Northeastern University Press.
——(1990) *Texts, Facts, and Femininity: Exploring the Relations of Ruling*, London: Routledge.
Smith, D. and Sage, M. (1985) 'Microcomputers and education in the United Kingdom: towards a framework for research', in Duncan, K. and Harris, D. (eds) *Computers in Education*, Proc. IFIP TC3, 4th World Conference on Computers in Education, North Holland: Elsevier.
Sokolowski, R. (1988) 'Natural and artificial intelligence', in Graubard, S. (ed.) *The Artificial Intelligence Debate*, London: MIT Press.
Spivak, G. 'Reflections on cultural studies in the post-colonial conjuncture', *Critical Studies: Cultural Studies Crossing Borders*, 3:1, 63–78.
Spufford, M. (1981) *Small Books and Pleasant Histories*, Cambridge: Cambridge University Press.
Stanley, L. (1991) 'Feminist Auto/Biography and Feminist Epistemology', in Aaron, J. and Walby, S. (eds) *Out of the Margins: Women's Studies in the Nineties*, London: The Falmer Press.
——(1994) 'The knowing because experiencing subject: narratives, lives and autobiography', in Lennon, K. and Whitford, M. (eds) *Knowing the Difference: Feminist Perspectives in Epistemology*, London: Routledge.
Stigler, S. (1986) *The History of Statistics*, London: Belknap Press.
Stock, B. (1983) *The Implications of Literacy: Written Language and Models of Interpretation in the Eleventh and Twelfth Centuries*, Ithaca NY: Princeton University Press.
Stockinger, P. (1990) 'Logicist analysis and conceptual inferences', in Ennals, R. and Gardin, J.-C. (eds) *Interpretation in the Humanities: Perspectives from Artificial Intelligence*, British Library, Library and Information Research Report 71, London: British Library.
Streuver, N. (1983) 'Lorenzo Valla: humanist rhetoric and the critique of the classical language of morality', in Murphy, J. J. (ed.) *Renaissance Eloquence*, London: University of California Press.
Strong, T. (1993) 'How to write scripture: words, authority, and politics in Thomas Hobbes', *Critical Inquiry*, Autumn.
Swearingen, C. J. (1991) *Rhetoric and Irony: Western Literacy and Western Lies*, Oxford: Oxford University Press.
Tanesini, A. (1994) 'Whose language?', in Lennon, K. and Whitford, M. (eds) *Knowing the Difference: Feminist Perspectives in Epistemology*, London: Routledge.
Tilghman, B. (1987) 'Seeing and seeing-as', in Goranzon, B. and Florin, M. (eds) (1990) *Artificial Intelligence, Culture and Language: On Education and Work*, London: Springer Verlag.
Tulviste, P. (1979) 'On the origins of theoristic syllogistic reasoning in culture and the child', *Quarterly Newsletter of Comprehensive Human Cognition*, 1.
Turkle, S. (1984) *The Second Self: Computers and the Human Spirit*, London: Granada.
——(1988) 'Artificial intelligence and psycholanalysis: a new alliance', in Graubard, S. (ed.) *The Artificial Intelligence Debate*, London: MIT Press.
Turkle, S. and Papert, S. (1990) 'Epistemological pluralism: styles and voices within the computer culture', *Signs*, 16:1, autumn.
Turner, R. (1984) *Logics for Artificial intelligence*, Chichester: Wiley.

Violi, P. (1992) 'Gender, subjectivity and language', in Bock, G. and James, S. (eds) *Beyond Equality and Difference: Citizenship, Feminist Politics and Female Subjectivity*, London: Routledge.

Wajcman, J. (1991) *Feminism Confronts Technology*, Cambridge: Polity Press.

Wainwright, H. (1994) *Arguments for a New Left: Answering the Free Market Right*, Oxford: Blackwell.

Wagman, M. (1991) *Artificial Intelligence and Human Cognition: A Theoretical Intercomparison of Two Realisms of Intellect*, London: Praeger.

Wallace, W. (1989) 'Aristotelian science and rhetoric in transition: the Middle Ages and the Renaissance', *Rhetorica*, VII:1, winter.

Waltz, D. (1988) 'The prospects for building truly intelligent machines', in Graubard, S. (ed.) *The Artificial Intelligence Debate*, London: MIT Press.

Way, E. (1994) *Knowledge, Representation and Metaphor*, Oxford: Intellect Books.

Weizenbaum, J. (1976) *Computer Power and Human Reason: From Judgement to Calculation*, Harmondsworth: Penguin.

West, L. and Pines, A. (eds) (1985) *Cognitive Structure and Conceptual Change*, London: Academic Press.

Wilcock, J. (1990) 'A critique of expert systems, and their past and present use in archaeology', in Ennals, R. and Gardin, J.-C. (eds) *Interpretation in the Humanities: Perspectives from Artificial Intelligence*, British Library, Library and Information Research Report 71, London: British Library.

Williams, C. (1994) 'Feminism, subjectivity and psychoanalysis: towards a (corpo)real knowledge', in Lennon, K. and Whitford, M. (eds) *Knowing the Difference: Feminist Perspectives in Epistemology*, London: Routledge.

Winograd, T. and Flores, F. (1986) *Understanding Computers and Cognition: A New Foundation for Design*, Norwood NJ: Addison Wesley.

Winston, P. (1984) [1977] *Artificial Intelligence*, London: Addison Wesley.

Wittgenstein, L. (1952) *The Blue and Brown Books*, Oxford: Blackwell.

——(1967) *Philosophical Investigations*, trans. G. Anscombe, Oxford: Blackwell.

Wolff, J. (1990) *Feminine Sentences: Essays on Women and Culture*, Cambridge: Polity Press.

Women and Words Collective (c. 1985) *In the Feminine: Women and Words/Les Femmes et les Mots*, Edmonton: Longspoon.

——(1992) *Women into Computing: Teaching Computing, Content and Methods*, Keele: Keele University Press.

Woodhouse, D. (1985) 'Course integration', in Duncan, K. and Harris, D. (eds) *Computers in Education*, Proc. IFIP TC3, 4th World Conference on Computers in Education, North Holland: Elsevier.

Woodward, K. (1980a) 'Preface: the Devil's hand', in Woodward, K. (ed.) *The Myths of Information: Technology and Post-industrial Culture*, London: Routledge and Kegan Paul.

Woodward, K. (ed.) (1980b) *The Myths of Information: Technology and Post-industrial Culture*, London: Routledge and Kegan Paul.

Woolgar, S. (1989a) 'The ideology of representation and the role of the agent', in Lawson, H. and Appignanesi, L. (eds) *Dismantling Truth: Reality in the Postmodern World*, London: Weidenfeld and Nicolson.

——(1989b) 'Why not a sociology of machine? An evaluation of prespects for an asociation between sociology and artificial intelligence', in Murray, L. and

Richardson, J. (eds) *Intelligent Systems in a Human Context: Development, Implications and Applications*, London: Oxford University Press.

——(1991) 'The turn to technology in social studies of science', *Science, Technology and Human Values*, 16:1, winter.

Wright, R. and Lickorish, A. (1989) 'The influence of discourse structure on display and navigation in hypertexts' in Williams, N. and Holt, P. (eds) *Computers and Writing*, Oxford: Intellect Books.

Young, R. (1989) 'Human interface aspects of expert systems', in Murray, L. and Richardson, J. (eds) *Intelligent Systems in a Human Context: Development, Implications and Applications*, London: Oxford University Press.

Yuval-Davis, N. (1997) *Gender and Nation*, London: Sage.

Ziman, J. (1982) 'Basic principles', in Rotblat, J. (ed.) *Scientists, the Arms Race and Disarmament*, London: Taylor and Francis.

Index

Printed in the United States
by Baker & Taylor Publisher Services